Individual
and
Family Living
in Canada

Individual
and
Family Living
in Canada

Eva Meriorg

Maureen Holloway

IRWIN PUBLISHING

Toronto/Vancouver, Canada

Copyright 2001 by Irwin Publishing Ltd.

National Library of Canada Cataloguing in Publication Data

Meriorg, Eva
    Individual and family living in Canada

Includes index.
ISBN 07725-2901-9

1. Family –Juvenile literature. 2. Family life education-Juvenile literature. I. Holloway, Maureen II. Title.

HQ10.5.C3M47 2001   306.85   C2001-900904-6

Project Team
Project Manager and Developmental Editor:
Marilyn Wilson/Marwil Communications Inc.
Copy Editor: Shirley Tessier
Photo Researcher: Lisa Brant
Production Editor: Jennifer Howse
Cover Design: Dave Murphy/ArtPlus Ltd.
Text Design: Leanne O'Brien, Sandra Sled/ArtPlus Ltd.
Page Layout: Barb Neri, Leanne O'Brien/ArtPlus Ltd.
Illustration: Donna Guilfoyle, Jeremy Kessler/ArtPlus Ltd.

Thomson Nelson
1120 Birchmount Road
Toronto, On
M1K 5G4
www.nelson.com

ISBN-13: 978-0-7725-2901-5
ISBN-10: 0-7725-2901-9

Printed in Canada

Author Acknowledgements

Thanks to my husband, Garth, and my two daughters, Claire and Kate, for insisting that writing this book is a family responsibility to which they will contribute tea, dinner, and housework. Many thanks also to the thousands of students whose names and stories are woven into wonderful memories and who, along with my daughters, inspired the examples in this book.

Maureen Holloway

Thanks to all of my immediate and extended family for their support and inspiration. Special thanks to my husband, Eamonn for taking care of me and everyone else while I was writing and to my children, Eadonn and Jordanna, who bring me joy on a daily basis.

Eva Meriorg

The authors and publishers would like to thank the following people:

Advisory Committee
Jane Antal, The York Region District School Board
Pat Durst, formerly York Region District School Board
Noreen McCormick, Halton Roman Catholic Separate School District
Maria McLellan, Halton School Board
Donna Naylor, Gravenhurst High School
Donna Pree, Niagara Catholic School Board
Ann O'Donnell-Beckwith, Waterloo Catholic District School Board
Michelyn Putignano, Hamilton–Wentworth School Board
Suzanne Robertson, Dufferin-Peel Catholic District School Board
Bruce Rodrigues, Waterloo Catholic School Board
Carole Whelan, Toronto District School Board
Gergie Yerxa, Toronto District School Board

Table of Contents

CHAPTER

1

Starting Out! Individual and Family Living in Canada

What will you be studying in this course? What types of activities will you be experiencing? How can you be successful in your studies in this course? How does the study of individual and family living fit within social science? And who are all these people? You wonder, "Will I have to work with all of them? Will I have anything in common with them? Will we get along?"

Welcome to a new course and a new textbook! You are starting out on an exciting journey toward understanding your own life.

By the end of this chapter, you will be able to

* prepare simple research questions for studying individuals and families;

* identify social science methods used to obtain information from various media, technology, and human resources in order to examine aspects of individual and family living;

* record information and key ideas from your research and document the sources accurately; and

* work collaboratively with others to achieve learning goals.

You will also use these important terms:

anthropology

data

economics

facts

opinions

psychology

research question

sociology

social sciences

Your Journey into the Social Sciences

You are starting out on an exciting journey to examine your own life and the roles you play in your family and your community. Begin the study of your life by taking a good look at yourself.

The answers to questions you ask yourself will help you to define who you are at this time on your journey into adulthood.

Why study the family? You have been born into a family, and you know yours intimately. You have had opportunities to observe other families closely. Studying families will give you an opportunity to examine them objectively, setting aside your emotions as you analyse evidence. You will learn some things that support what you already know. More important, you will discover the truth behind some of the common beliefs about families. Since you will probably form a family of your own some day, knowledge of family living can help you to make that family what you want it to be. The benefits of a knowledge of family living can extend beyond your personal life if you choose to volunteer or work professionally with families in the future.

What Are Social Sciences?

Questions relating to the study of individuals, families, and their lives can take many perspectives. The ***social sciences***, which study human behaviour, include several subjects.

- ***Psychology*** examines how you think, how you feel, and what motivates you.
- ***Sociology*** studies how people behave with others within families, groups, and society.
- ***Anthropology*** studies humans and their cultures.
- ***Economics*** studies the production and distribution of wealth.

In family studies, as well as in other courses such as Challenge and Change in Society, you will use various social sciences to examine how you and your peers behave within your primary social group, the family.

The social sciences differ from other sciences. In physics, you will learn that an object that is dropped will always fall down, and in chemistry, you will learn that water is always formed by the combination of hydrogen and oxygen. In mathematics, when you add two plus two, the answer is always four. These are facts. While a falling apple cannot control its behaviour, people have the ability to make choices.

Social science research looks for the patterns in human behaviour as well as the connections among behaviours. Therefore, in the social sciences, you will learn about patterns of behaviour, not rules. Results of social science research are often expressed as percentages or ratios. You will read that 85 percent of teenagers identify their friendships as very important in their lives and

Each man must look to himself to teach him the meaning of life. It is not something discovered; it is something moulded.

Antoine de Saint-Exupéry

The object of psychology is to give us a totally different idea of the things we know best.

Paul Valery

that one in ten Canadian families composed of couples with children is a step family. A knowledge of social science enables you to predict what is likely to happen based on past experience, but you and everyone else could choose to be different.

In social sciences, the most frequently asked questions are "What happens?" and "Why?" You will try to answer these questions by learning and using the methods used by social scientists. Applying social science research methods will enable you to conduct your own research into the beliefs and behaviours of individuals and families. Based on the results of your studies you will be able to distinguish between personal opinions and facts. **Opinions** are based on individual observations or beliefs whereas ***facts*** are supported by evidence that everyone can observe. You will determine whether you have answered your own questions. You will also consider how what you have learned applies to your own life.

career Link

What is a social scientist?

Social scientists use social science research methods to learn about human behaviour.

The results of their studies may be published in academic journals for other social scientists to read and use as the beginning for their own research. These studies are also reported in the news, raising awareness of new knowledge about individuals, families, and their lives. The results can be used to make changes and improvements in our personal lives. The results are used by governments to plan social policy, by organizations to develop programs for people, and by business to develop new products and to sell goods and services using advertising. Watch for examples of social science research and ask how the results will affect your life.

Developing a Theory

When you try to answer the question "Why?" you are developing a theory. A theory is an attempt to explain patterns of behaviour. You develop theories all the time.

- **Gather evidence.**
  A good theory is one that is based on evidence such as facts, testimony, or documentation.

- **Make predictions.**
  The evidence helps you to predict fairly accurately what will happen next time.

  For example, a principal develops a theory that three times as many students skip classes on the first warm sunny day in spring as on any other day. The theory is a good theory if she has checked the attendance records against weather records for evidence, and if the attendance for the first warm sunny spring day next year matches her predictions. Because there could be many theories to explain a particular behaviour, discussion and debate are the focus of the social sciences.

It is the theory that determines what we can observe.

Albert Einstein

Once you understand social science research methods, you will be able to read and analyse the results of the research done by others in order to increase your knowledge and understanding of individuals, families, and communities. You will be able to look for the reasons behind social policies, laws, and the types of programs available. You will also be able to analyse the purposes of advertising. As a result, you will gain a better understanding of the many influences on your life.

Social Science Research Method

Social science research is carried out step by step. Often the method is straightforward, but sometimes it is necessary to go back and revise earlier steps as your study progresses in unexpected ways.

- **Identify the problem.**
  Begin with a topic or issue you want to study. Determine what you already know about the topic. The problem states what you need to find out.

- **Formulate a research question.**
  A research question forms the basis for your investigation. A good *research question* states exactly what you want to learn and will suggest how you will conduct your research. Experienced researchers develop a hypothesis—a statement of a possible answer to the question—which they will attempt to prove or disprove through research.

- **Gather data.**
  Gather information, called *data*, to answer your research question. Methods of gathering data include sample surveys, such as questionnaires or interviews, experiments, and observations.

  It is important to distinguish between *fact* and *opinion*. Facts can be supported by evidence, but opinions are personal perceptions.

- **Analyse the data.**
  Organize your data so that you can compare, analyse, and summarize the data. Spreadsheets, tables, and graphs might be useful. Look for the relationships between your data and your research question.

- **Form and communicate conclusions.**
  Form conclusions that state how the data answers your question. You should also draw conclusions about the research method you have used. Communicate your conclusions as a written, oral, or visual research report.

CHECK ✓ Points

These are questions that check that you have understood what you have read. Use them
- as an outline for note making,
- as prereading questions, and
- as review questions.

- ✓ What are the disciplines that contribute to social science?
- ✓ What is the focus of each of the disciplines?
- ✓ What are the reasons for family studies?
- ✓ What are the steps in the social science research method?

Gathering data using interviews will help you distinguish between fact and opinion.

Expand Your Learning Skills

Success in your studies depends on how well you develop the skills required at school. What are you expected to do in school? What skills do you have already? What do you need to improve? When you understand how learning occurs, you can become better at your school work.

Steps in Knowing

Learning is a multistep process that occurs over time. Knowledge does not exist in tidy categories that fit neatly between the covers of books such as this one. Knowledge, like life, flows together seamlessly, constantly changing and evolving. Schools sort knowledge and skills into "subjects," so do not be surprised by the overlaps among your various subjects at school. Look for the connections among the subjects and within your life. These connections will provide you with the meaning of what you are learning. Understanding the meaning of what you have learned will enable you to use what is relevant to solve problems in the real world. Understanding takes time. Being able to apply your knowledge and skills takes practice.

Application
- Use your knowledge and skills in familiar and new situations in the classroom and in real life.

Communication
- Describe and explain your new knowledge and skills to others using specific language and a variety of verbal and visual methods.

Thinking and Inquiry
- Use inquiry/research skills, and critical and creative thinking skills to enhance your knowledge and skills.

Knowledge and Understanding
- Know and understand new facts, theories and skills and understand how they are related to what you already know.

Reading Your Textbook

As you begin to read this textbook, notice how it is organized to enable you to learn effectively. Your textbook is divided into units of study or chapters. Each chapter includes:

- an **overview** to introduce you to the questions you will be answering as you read;

- **expectations** that tell you what you will be able to do by the end of the chapter;
- **important terms** that identify the vocabulary necessary for understanding the chapter contents; and
- **headings** that identify the chapter topics and can be used as an outline for making notes.

After you have read a section in the textbook, use the features provided to help you understand and apply what you have learned.

- **Check Points** are questions which can be used to review the material you have learned and check your understanding.
- **Reflections and Connections** provide opportunities to apply what you have learned to your own life, to look for connections with what you already know, and to discuss the meaning of what you have learned with others.
- **Quotations** provide alternative viewpoints for reflection and class discussions.
- A **Summary** at the end of each chapter identifies the main points of what you have read and might help you to highlight the key ideas from the chapter.
- **Activities to Demonstrate Your Learning** suggest assessment of the expectations.
- **Enrichment** activities suggest possible connections to other disciplines, further study, and independent study.
- **Hands On** sections of the text are designed to have students apply their learning to practical situations.

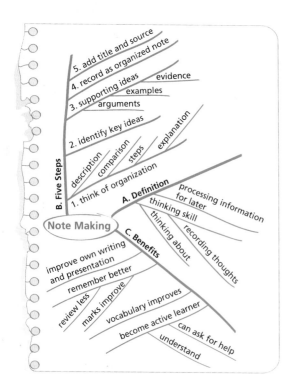

Note Making

Note making is actually a thinking skill that allows you to do two things at the same time. While part of you is thinking about what you are reading or hearing, the other part of you is recording your thoughts for review later. You are processing information and storing it in a manageable form.

There are five steps in making notes, whether you are reading or listening. First, determine whether the material is organized as a description or a comparison of characteristics, a list of steps or ideas, or an explanation of arguments or reasons. This will enable you to understand how the ideas will be connected. Secondly, identify what is important, the key ideas. Learn how to get to the point of what you are reading or hearing. Next, identify the supporting ideas, the arguments, examples, or evidence presented to support the key ideas, and consider whether they are effective. As you locate these three levels of ideas, the next step is to record your thinking briefly in your own words in an organized note. The note should distinguish between key ideas and supporting ideas, and show how the ideas are related to each other. As a final step, remember to add a title. Use the headings from the text, as well as your own key words in the title, to clearly indicate the contents of your note. Do not forget to include a reference to the source of the information, even if it came from your textbook. Developing note-making skills will help you in several ways. Processing the information into your own words as you make notes makes you an active learner. You will understand better, and you will know when to ask for some explanation of the ideas you do not understand. Your vocabulary will improve. You will remember what you have learned better, you will need to review less, and your marks will improve. Analysing how others present information when you are making notes will improve your own ability to present your ideas in essays or oral presentations. Develop your note-making skills and you will reap the benefits!

The work will teach you how to do it.

Estonian proverb

Working with Others

How do you prefer to work? Are you happiest working quietly by yourself? Do you prefer to work with one other person? Do you enjoy being part of a busy group of several people, each carrying out part of the task? In class discussion, do you sit quietly listening to other people, do you "tune out," hoping the teacher will not call on you to answer, or do you want to answer every question? Working collaboratively with others means that you can contribute to group activities which make learning quicker and easier for all. As they say, "Two heads are better than one," and "None of us is as smart as all of us!"

You are beginning a study of individual and family living. You will learn about your life as a teenager, where you fit into your family, your school, and your community. You will learn how to relate to others, and how to manage the many resources available to help you make the best of your life. You will develop many learning skills along the way. Good luck! Bon voyage!

Working collaboratively with others can enable everyone to be more successful.

Collaborative Group Behaviour

1. Perform an effective role in the group.
- Resources managers determine the needs, then locate and obtain the materials needed to complete the task.
- Information managers lead groups in the sharing and discussion of ideas and keep records.
- Product managers clarify the task to be achieved and ensure that it is completed to a high standard.
- Social managers make sure that everyone is involved in the task and support and encourage participation.
- Timekeepers determine the time lines and ensure that the group stays on task to meet the deadlines.

2. Share the workload.
- All aspects of the project are divided so that members are doing a fair share of the workload, although they might not do the same work. Every member needs to participate in the creation of the solution to the task.

3. Share and discuss ideas.
- All group members share their ideas with the group, invite others to share their ideas, and listen carefully to others. Decisions are discussed until everyone agrees.

4. Stay on task.
- Effective group members have a clear understanding of the group goal and stay focused so that the final product is well done according to the expectations.

Reflections and Connections

The questions in these features ask you to think about how what you have learned relates to your life. Use them

- to reflect on your personal thoughts and feelings for examples of what you have learned;
- to make connections between your experiences and those of others; and
- to share your experiences as examples to help others understand what they are learning.

1. What are you hoping to learn in this class? Compare your goals with those of your classmates.
2. What skills do you have that will be useful in this study? What do you need to learn to do?
3. How do you think you will change during this course? Are your feelings shared by your classmates?

Summary

1. The study of individual and family living is a social science, the study of human behaviour.

2. The social sciences include the disciplines of political science, psychology, sociology, anthropology, and economics.

3. Family studies, and other similar courses apply the various disciplines of social science to the study of how people behave within their primary social unit—the family.

4. By applying social science research methods, it is possible to form conclusions about questions related to individual and family living.

5. Skills such as thinking and reflecting, note making, research, and collaborating in groups are necessary for the study of individual and family living.

In this chapter, you have learned to

- prepare simple research questions as the basis for studying individuals and families;
- identify social science methods used to obtain information from various media, technology, and human resources in order to examine aspects of individual and family living;
- record information and key ideas from your research and document the sources accurately;
- work collaboratively with others to achieve learning goals.

Activities to Demonstrate Your Learning

1. Take a photograph of yourself and use it as the foundation of a collage entitled, "This Is Me!" Use words, images, and artifacts around the photograph to represent you as you are today. Include ideas about what you look like, how you feel, what you think about, what is important to you, what you like to do, your pet peeve, the important people in your life, what you can do well, the challenges in your life, and how you spend your money. Post the collages on the bulletin board so that you can get to know your classmates.

2. In collaborative groups, use social science research methods to find out what issues concern teenagers. Report your conclusions as bar graphs to be posted in your classroom.

3. Find an article in a newspaper or magazine or on the Internet about one of the issues that concern teenagers. Use note-making skills to summarize the key ideas in the article, and place the article in an issues file. Share one or two of the key ideas with others in small groups. Discuss the issues and select the two most important ones.

4. Write a letter to yourself to be read at the end of this course. Reflect on your thoughts and feelings at this time in your life. Outline your hopes and dreams for the coming year. Describe the most important issue in your life right now and how you hope it will turn out. Seal and save the letter.

CHAPTER

2

Life Is a Journey!

Your journey through life begins at your birth and ends at your death. What lies ahead? Your journey is predictable in many ways, but your trip will be unique, a sum of your personal life experiences. What surprises will there be around each bend? Will there be long stretches of road when nothing seems to change? Will you make a turn that provides opportunities that change your life forever? Or will you have to turn back and try another direction? Your life does not come with clear directions, or even with a road map, so you must decide which direction to go. What an adventure!

By the end of this chapter, you will be able to

❋ outline key aspects of your development in adolescence;

❋ summarize how your development affects your behaviour;

❋ explain how your needs change as you develop; and

❋ explain the influence of heredity and environment on your development.

You will also use these important terms:

adolescence	environment
adolescents	growth
development	heredity
developmental tasks	life span

Lifeline

0–1 year

Infancy

1–3 years

Toddler

3–5 years

Preschool

5–12 years

School-age

12–18 years

Adolescence

Your growth and development will occur in a common pattern but at your own pace.

The Journey of Human Development

Your journey through life began when you were born. Imagine that you are looking through your family photo album. You might see important events of your life that were recorded, such as your birthdays, your first steps, or the first fish you caught. Perhaps your family has saved mementos of your childhood, such as a lock of hair from your first haircut, the program from a school concert, your report cards, etc. The events you have experienced so far may be similar to those of your classmates, but there will be many features unique to your journey.

Your memory is your record of the significant experiences of your life. What have been the three most important events in your life so far? You can probably replay the sights, sounds, and even the feelings of a family trip, your first crush, or the applause after a successful presentation at school. Every experience, both good and bad, leaves its mark on you. If the events have been mostly positive, you will be excited about continuing on to the next stage in life. Many negative events can hamper your ability to move on to new challenges. These important marker events stand out as milestones on your journey through life.

Patterns of Development

Your journey through your life requires growth and development at every step along the way. During your physical growth, your bones got longer, your muscles became stronger, and your body proportions changed. Did you know that your brain cells also grew rapidly in the first years of life? Growth and development occur in several areas, not only physical, but intellectual, emotional, and social as well. ***Growth*** refers to the physical changes that occur in the body. ***Development*** occurs when you co-ordinate skills into complex behaviours. Growth in your leg bones and muscles was necessary for you to continue to be able to support your own weight; co-ordination of several movements into a complex pattern was necessary for you to develop the skill of walking.

There are predictable stages of growth and development. Social scientists such as Erik Erikson have identified stages of the human life span and the patterns of growth and development at each stage.

The journey begins with infancy. It is followed by childhood that includes the toddler, preschool, and school-age stages. You are currently in adolescence. You will move through early adulthood, middle adulthood, and the aging years before your journey ends. Each stage presents you with challenges for growth and development in several areas.

Growth and development occur in an orderly sequence. Normal development includes crawling before walking, and walking before running. Each individual grows and develops in the same order, but at a unique rate.

Matthew, who was taller than his classmates at twelve, will not necessarily be the tallest at high school graduation. Early intellectual development enabled Frances to be top of the class in Grade 9, but this does not necessarily mean that she will be more intelligent in adulthood. Your rate of development does not limit your development; however, each stage of your journey has an impact on the next stage.

Adolescence is the stage of life that begins around puberty. Adolescence as a separate stage of development is fairly new—people used to be considered children until they became adults. Now, adolescence is considered by some people to continue to about 20 years, or even longer. Individuals in this stage are called adolescents or, more commonly, teenagers. The journey through adolescence is a journey of self-discovery. *Adolescents* clarify who they are, what they can do, how they feel, what they believe and where they fit in. Adolescence is an important transition in your life as you complete your childhood development and prepare for the journey into adulthood.

Developmental Life Tasks

Stage of Life　　　　　　　**Developmental Life Task**

Infancy

Trust vs. Mistrust
Children learn to trust others and feel secure that the world is a safe and predictable place.

Toddler

Autonomy vs. Doubt
Children learn to do things with minds of their own and to act independently.

Preschool

Initiative vs. Guilt
Children learn to act with a sense of purpose and to try again if they are not successful.

School-age

Industry vs. Inferiority
Children learn to work hard to develop their skills and talents and to be proud of their individual accomplishments.

Adolescence

Identity vs. Role Confusion
Adolescents develop a sense of who they are, what they can do, and where they are going.

Early
Middle—**Adulthood**
Aging

career Link

What is a psychologist?

A psychologist studies the processes of the brain and human behaviour from a scientific point of view in order to help people understand, explain, and change behaviour.

What do psychologists do?
Psychologists study the biological and social principles of learning, behaviour, emotion, and human interaction with environmental, biochemical, and genetic factors. They do this in order to answer research questions about various aspects of people's lives. They use this information to help people by designing therapeutic interventions that assist in behaviour changes for individuals and groups.

Where do psychologists work?
Psychologists work in a variety of community settings such as hospitals, rehabilitation centres, and schools. They also work in business and industry as well as private practice.

Life is not a destination, but a journey.

Unknown

TO DO BEFORE YOU REACH THE AGE OF 20!

The Developmental Tasks of Adolescence

■ **Understand and accept who you are.**
- Learn about your personality, traits, and your skills.
- Identify and honour your strengths and weaknesses.
- Develop and maintain a healthy body image.
- Clarify your values, beliefs, and standards.

■ **Make effective choices that help you mature.**
- Identify realistic goals for yourself.
- Develop and use decision-making strategies.
- Make effective use of your resources.

■ **Develop mature relationships with others.**
- Develop and use effective communication skills.
- Express and respond to emotions appropriately.
- Learn the importance of commitment in a variety of relationships.
- Learn to co-operate and resolve conflict.

■ **Achieve responsibility and independence.**
- Become autonomous by acting on your own.
- Learn to accept the consequences of your actions.
- Consider the influence of your behaviour on other people.
- Take on meaningful roles in your family and your community.

■ **Prepare for your career.**
- Determine your interests, talents, and skills.
- Acquire and develop skills needed to achieve your goals.
- Explore career opportunities and pathways.

CHECK ✓ Points

1. ✓ List the stages of the human *life span*.
2. ✓ What is the difference between growth and development?
3. ✓ Distinguish between adolescent and adolescence.
4. ✓ What is the major challenge of adolescence?
5. ✓ What are the developmental tasks of adolescence?

Each stage of life presents challenges for growth and development called ***developmental tasks***. The psychologist Erik Erikson defined the major challenge of adolescence as developing an understanding of who you are and will be, and the role you will play in life. He believed that adolescents must develop an identity or risk confusion about their role. In order to develop an identity, you will acquire mature skills, habits, knowledge, and attitudes. The specific developmental tasks of adolescence define in greater detail how you will face the challenge of developing an identity. You will be studying these tasks in later chapters. How well you manage the developmental tasks on your journey through adolescence determines your readiness for adulthood.

Reflections and Connections

1. Describe ways in which individuals you know have developed differently from others.
2. List other terms that are used for "adolescent." Discuss their connotations; that is, how they imply positive and negative attitudes towards adolescents.
3. Identify the developmental tasks that you think might be the most challenging and explain your choices.
4. Describe how other people have influenced your development in childhood and early adolescence.

Areas of Development

When you look closely at the photographs of yourself, your family, and your friends, it is easy to see the signs of physical growth and development. Significant changes occur in other areas at each stage of life as you develop intellectually, emotionally, and socially. Development in these areas results in changes in behaviour. Behavioural changes are more difficult to recognize than physical changes, so you might not anticipate them as easily.

Physical Development

Your physical development in adolescence began at puberty. Your pituitary gland produced a growth hormone so that your body began to take on its adult size and proportions. The increase in the levels of sex hormones—estrogen for females and testosterone for males—resulted in the development of the secondary sex characteristics—the outer characteristics that signal your sexual maturity and result in your distinctive male or female appearance. At the same time, your sex organs reached full size, and as they began to work, you became fertile. You will continue to grow for a few more years until the last part of your body, your feet, stop growing. Then you will see what your final adult body will look like.

Some teens feel uncomfortable with their changing body image. If you mature sooner or later than others, you might feel self-conscious. You might be surprised by the changes in your hair colour, your facial features, or your body shape. Many teenagers are surprised to discover how much, or how little, they resemble their parents. Some teens, both boys and girls, focus on their so-called flaws when they compare themselves with the images presented by the media. Because genetics plays the greatest role in your physical growth, most of it is beyond your control. These confusing feelings about physical growth are common in adolescence.

Physical development is easily measured; your intellectual, emotional, and social development are less visible.

As your body approaches its adult size and proportions, you can develop and use your strengths and skills.

Once you become accustomed to these physical changes and developments, you will be able to develop and use your physical strengths and skills. You will also take on the responsibility for the care and maintenance of your body. The key to successful physical development is to recognize that you cannot control growth, but you can affect the development of your skills. Be realistic about your needs. Healthy eating habits, adequate sleep, regular exercise, and good grooming will help you make the most of your body. Choosing to participate in physical activities will enable you to develop your physical skills. Your appearance and performance reflect your growth, your development, your family, and your heritage. You are really unique!

Intellectual Development

Your brain had a growth spurt recently in which many more connections were made between your brain cells. You might not have noticed because your hat still fits, but as a result of this growth, you are developing new intellectual skills. As a child you were only capable of concrete thinking based on observing the real world. You have studied changes of state in science. As a child you could understand the change from solid to liquid only if you saw ice melting into water. In adolescence, you are developing skills for abstract thinking. Thinking about ideas that cannot be seen such as love, truth, or fairness—abstract thinking—challenges you to develop more complex thinking skills. As your intellectual skills develop, you will be thinking more about the meaning of the events and relationships in your own life and in society.

Adolescence is the time when logical thinking skills develop. As you become skilled at connecting ideas, you will develop your ability to draw several ideas together in order to solve more complex problems. Your ideas, opinions, and thinking will develop as you are exposed to new ideas. You will also become more skilled at predicting possible consequences of your choices when you are making decisions. Your creative thinking skills will also continue to develop as you find new opportunities to use them in your day-to-day activities. These combined skills will enable you to make decisions and solve problems. They will also help you to plan ahead and take responsibility for your life.

To develop your intellectual skills you need to exercise your brain. Your teachers help you develop your thinking skills when they ask you difficult questions and set challenging assignments. A variety of activities in school enable students to learn according to their personal learning styles and to strengthen multiple intelligences. Participating in activities at school and in the community gives you

CHECK Points

✓ What determines your adult size and proportions?

✓ What are some of the primary and secondary sex characteristics which develop in adolescence?

✓ Why should you not invest in a lifetime supply of shoes now?

✓ Why do some people feel uncomfortable with their physical growth?

✓ What steps can you take to develop your physical strengths and skills?

Teenagers' brains really are different from their parents'

Hot topic at conference on frontal lobes

By Tanya Talaga
MEDICAL REPORTER

The frontal lobes of the brain keep changing as humans enter early adulthood and perhaps beyond, influencing everything from how we learn to our social behaviour and the people we'll become as adults.

For years, researchers have thought that our genetic load and our environment shapes who we are, said Dr. Robert Knight, a professor in the department of psychology at the University of California.

But it turns out things are a little more complicated than that, said Knight, who is co-chair of the Rotman Research Institute's 10th annual conference, being held this week in Toronto.

"There are waves of development, up to at least the age of 22, that all involve changes in the prefrontal cortex," he said.

"It turns out there are huge changes between the ages of 13 and 22."

Recent research shows that a teenager's brain doesn't just function differently, it actually is structurally different from that of a fully formed adult.

Frontal lobes are conductors for symphony of brains

This "incredibly hot area" of brain research will be discussed later in the week, said Knight, on of the over 800 top neuroscientists from 32 countries who have come to Toronto for the conference.

This year's theme is the frontal lobes, one of the most intriguing and least understood areas of the human brain.

How important are the frontal lobes?

Think of the brain as a symphony, says Knight.

All parts of the brain are different instruments that play beautifully on their own.

"But if there isn't a conductor to make it a coherent orchestra, it's just going to be a bunch of noise," he said.

Unlocking the mysteries of the brain may not be as popular as the human genome project – the global race to create a blueprint of the human genetic code – but it's equally important, said Donald Stuss, director of the Rotman institute, which is affiliated with the Baycrest Centre for Geriatric Care.

The frontal lobes function as the higher-thinking parts of the brain and are what make us human, said Stuss.

The lobes control everything from empathy to social interaction, judgment and the integration of emotion and thinking.

Scientists' understanding of certain diseases, such as dementia and Alzheimer's disease, is improving because of advances in our understanding of the frontal lobes.

The Toronto Star, March 21, 2000 Reprinted with permission—The Toronto Star Syndicate

You are becoming more capable of considering all of the factors when you are attempting to solve a problem.

opportunities to make plans and solve problems. At home, taking on responsibilities for tasks means that you can help make choices and solve problems as they arise. Seeking opportunities to discuss ideas and argue your point of view on issues with family, friends, teachers, and classmates will give your brain the workout it needs to develop intellectual skills.

"Mens sana in corpore sano"
The motto of Dr. G.W. Williams Secondary School in Aurora, Ontario, means that it is important to have a strong and healthy mind as well as a healthy body.

Multiple intelligences

The work of Howard Gardner and other scientists has identified specific intelligences that are common to all human beings, and that vary in degree in each person. The multiple intelligences, ways of learning, and knowing are as follows:

Verbal/Linguistic Thinks and learns through written and spoken words; has the ability to memorize facts, fill in workbooks, take written tests, and enjoy reading

Logical/Mathematical Thinks deductively; deals with numbers and recognizes abstract patterns

Visual/Spatial Thinks in and visualizes images and pictures; has the ability to create graphic designs and communicate with diagrams and graphics

Body/Kinesthetic Learns through physical movement and body wisdom; has a sense of knowing through body memory

Musical/Rhythmic Recognizes tonal patterns and environmental sounds; learns through rhyme, rhythm, and repetition

Intrapersonal Enjoys and learns through self-reflection, metacognition, working alone; has an awareness of inner spiritual realities

Interpersonal Thinks empathetically; learns through social communication

Existential Is concerned with ultimate life issues—love, death, philosophy. Learns in context with meaning

Naturalist Loves nature and the out-of-doors. Enjoys classifying species—flora and fauna

Adapted from *Frames of Mind: Theory of Multiple Intelligences*, 1985

Emotional Development

Your emotions are your inner feelings. You experience and express your emotions in many ways. You probably need to feel secure to be able to trust people in your life, and you need to be accepted and loved by others. When you were a child, you trusted and liked almost everyone. You were generous with your smiles and hugs. However, your feelings were easily hurt and you cried easily. You expressed your feelings of independence by insisting, "Me do it!" By elementary school, you learned to look for opportunities to feel success and accomplishment. These feelings of trust, acceptance, independence, and competence still represent your major emotional needs.

CHECK Points

✓ Why do thinking skills develop in adolescence?

✓ Distinguish between concrete and abstract thinking.

✓ What kind of thinking skills are developing during this stage of your life?

✓ What are the different intelligences?

✓ How can you exercise your thinking skills?

The Big Eight Emotions

These are the universal emotions shared by all humans:

 Fear

 Sadness

 Anger

 Joy

 Surprise

 Disgust

 Anticipation

 Acceptance

All other emotions are combinations of these basic emotions
(e.g., surprise + sadness = disappointment; anticipation + joy = excitement)

Children express their emotions freely without embarrassment. They cannot think about how they look to others, nor can they consider the effects on other people. Think of a child throwing a temper tantrum because he is frustrated at the block tower falling! You are learning to recognize and understand your emotions better because of your newly-developed intellectual skills. You are more able to make judgments so that you can trust others and express your emotions in acceptable ways. As an adolescent, you are learning to use appropriate emotional reactions that do not hurt you, your reputation, or other people.

When you were a child your parents and other adults looked after your feelings. They also guided your emotional development by helping you to identify what you were feeling and correcting you when you expressed your feelings in an unacceptable way. Now you are learning to identify and meet your own needs. When a friend comments, "It sounds like you're feeling disappointed," she is helping you identify your feelings. If your friend's identification of your feelings is incorrect, you know you need to express your emotions more accurately. Accepting responsibility for the effects of your behaviour on others will also help you learn to control your emotional responses. You are finding out how to balance dependence and independence by expressing and discussing your feelings in order to manage them.

Above: Recognizing the negative emotions being expressed by others allows you to respond appropriately with sympathy.

Left: Emotional maturity means that you can celebrate the success of your friends.

Emotional
intelligences

According to Peter Salovey in Daniel Goleman's book, *Emotional Intelligence*, 1995, the five domains of emotional intelligence are

1. knowing one's emotions;
2. managing emotions;
3. motivating one's self;
4. recognizing emotions in others, and
5. handling relationships.

Emotional skills are so important that Daniel Goleman refers to them as emotional intelligences.

CHECK Points

✓ Identify situations which illustrate each of the basic emotions.
✓ Why is it necessary to learn how to express emotions?
✓ How do you learn to express emotions in a socially acceptable manner?

Social Development

On your journey through life, you will cross the paths of many people. Humans are interdependent; that means that you depend on and are influenced by people every day, and they in turn depend on and are influenced by you. Other people influence you if they give you a hug, give you food, teach you math, or pass a law changing the minimum wage. You influence others when you offer encouragement, vacuum the living room, or spend your money at the local mall. Effective interactions with other people are a sign of your social maturity.

Your social development helps you meet several needs. You need a sense of belonging. As you move from your family out into the world, your need to belong expands into a need to be accepted by society. Taking on meaningful roles means that you need to co-operate, work, and resolve conflicts with others. These social needs are met by developing the social skills necessary for effective relationships.

Children are naturally self-centred. In order to meet their own needs, they relate to others from their own point of view. You are learning to share viewpoints with others and think about other people's needs. You can help out your family members when they need your support. You can influence your friend by offering encouragement when he is afraid to try out for a school team. You can learn to become a team player who shares with others and accepts responsibility for your actions by participating in sports or clubs. You might also become a leader who accepts responsibility for others as a volunteer in the school or the community. Social development in adolescence enables you to co-operate and compromise with those around you.

Social development enables you to develop the relationships you need to feel accepted by others and to take on meaningful roles.

teen quiz

Quiz: Are You Socially Mature?

1. When you have made a commitment to someone to complete a task by a certain time you

   (a) ask your friend to keep reminding you about it;
   (b) exercise self-discipline by working on it for an hour each day until it done; or
   (c) put it off until the last day and pull an "all-nighter" to finish the task.

2. You have a regular Thursday evening baby-sitting job with a family where both parents attend night school. They depend on you to be there promptly at 6:00 P.M. Your friend has just called to invite you to see your favourite rock group—she just got tickets from her cousin who could not go. You desperately want go to the rock concert, but you know that the family you baby-sit for is depending on you. You decide to

   (a) go to your baby-sitting job;
   (b) call the family and tell them that something has come up and you can not baby-sit tonight; or
   (c) bribe your little sister to go baby-sitting for you.

3. You have received permission from the principal to use a classroom after school to hold a meeting with some other students. You have had to sign for the room on behalf of your student group. When the meeting is over, you see that the other students have left empty cups and plates all over the room and the seats are out of order. You decide to

   (a) find the caretaker and tell her about the mess;
   (b) clean the room up yourself; or
   (c) leave the room as it is—after all, you did not create any of the mess.

4. You strongly believe in working hard to achieve good grades. In the past you have been tempted to plagiarize on essays, but have stuck to using your own original work. You have never cheated because you really want to know that you have earned the marks that you get. Your friends have cheated in various ways and have offered you the opportunity to cheat. Up until now you have always refused. The final English exam is next week and you are not pleased with the marks you have earned so far in the semester. One of your classmates has stolen a copy of the exam from the photocopying room and is offering to sell you a copy. You decide to

   (a) buy a copy of the exam hoping to boost your mark;
   (b) refuse to buy the exam and study very hard; or
   (c) speak to some of the students who have bought a copy of the exam to see what hints you can get from them.

5. Science has never been your favourite subject, but passing it this year is required in order to achieve your high school diploma. You know that somehow you need to get motivated. You decide to

   (a) ask your parent to give you money if you pass;
   (b) go through the motions of attending class and see what happens; or
   (c) set the goal of a 75 percent grade as a personal challenge to work towards.

6. You have seen one of your classmates stealing money from someone else's locker. The person from whom the money is being stolen is really well-off and is always boasting about how much money he has. Some of your friends also know who the thief is, but because she seems to be a nice person and comes from a poor family the other students seem to think it is okay. You are asked by the vice-principal to give information about the situation. You decide to

   (a) report the facts as you have witnessed them;
   (b) keep quiet and say you know nothing about the situation; or
   (c) tell the vice-principal that you will talk to the thief and ask that they give the money back.

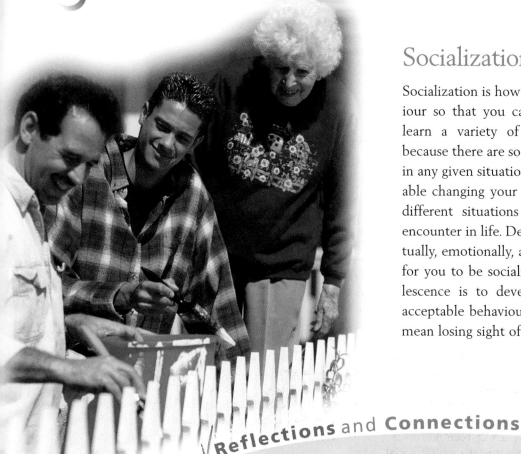

You can express your motivation and your responsibility by volunteering to help others.

Socialization

Socialization is how you learn acceptable behaviour so that you can fit into society. You will learn a variety of behaviours and responses because there are so many different expectations in any given situation. You will become comfortable changing your behaviour to suit the many different situations and experiences you will encounter in life. Developing physically, intellectually, emotionally, and socially are all necessary for you to be socialized. Your challenge in adolescence is to develop personal standards for acceptable behaviour, so that fitting in does not mean losing sight of who you are.

Reflections and Connections

1. Describe ways in which teenagers might be surprised or disappointed by their development.
2. Discuss how your family encourages your intellectual development.
3. Describe situations in which teenagers develop emotional and social skills.
4. In your opinion, does your community encourage adolescents to develop their social skills?
5. Identify skills you would like to develop. What steps are you taking to achieve them?

CHECK Points

✓ What is meant by interdependence?
✓ List and explain the basic social needs of humans.
✓ How and why are children self-centred?
✓ How can teenagers develop social maturity?

Your social development prepares you to take on a variety of mature responsibilities in your community.

FUTURE SHOCK TEENAGERS CARE ABOUT CAREERS

By Dr. Sandra Collins

What comes to mind when society thinks of teenagers? Some stereotypes of adolescence involve experimentation with drugs, preoccupation with sexuality, expressions of aggression, peer pressure, a sense of invincibility, and an almost complete absorption with the here-and-now.

What would make anyone think that teenagers would care about career development activities? Career development, after all, demands thinking about yourself, the future and your place in it. Teens couldn't possibly be interested in that, could they?

Perhaps the voices of adolescents themselves are what we need to pay attention to. According to a survey of almost 2,900 adolescents from five Calgary high schools, one of the things that they care most about is preparing themselves for their future work life. In this study, students were asked to identify their most pressing needs related to mental, emotional, physical, social, academic, and environmental well-being.

One of the results that stands out is the emphasis that both male and female adolescents placed on planning for the future. The number one type of counseling service these students felt they required was career counseling. In fact, in many cases, it was the only type of counseling that students felt was not being adequately addressed in schools. In a period when resources for health and guidance in schools are being continuously reduced, it is no wonder that teenagers sometimes feel like their voices are completely ignored.

The main theme that ran through all areas surveyed was the need for more proactive skill development.

Students wanted to learn how to be better problem-solvers, make effective decisions, handle their stress, improve their study skills, and manage their time and money more effectively, to name a few examples. One student in this study commented that this was the first time anyone had ever asked her what it was she thought she needed to learn. Maybe it is time we begin to rethink the image our culture has created around adolescents and listen to what they are actually trying to tell us.

One of the most disconcerting elements of this study was the degree to which major differences existed between what adolescents think they need and what their parents and teachers believe is important for them to develop. One would have expected adults to be more focused on helping students begin building for the future, but for the most part the adults surveyed, particularly school personnel, tended to be less future-focused and more problem-oriented than the students.

Perhaps it is time that we all shifted our focus away from the image created in the popular media of adolescence as a time of storm and stress, and focused our attention on what today's young people are actually trying to say.

The message seems clear: Invest in the future by investing in youth—what they want and need rather than what society is comfortable with giving them. Avoid future shock by helping our youth plan for their careers and futures…

Reprinted with permission of *The Edmonton Sun*

Heredity and Environment

You have many characteristics that make up your personality. As you continue to develop through life, you will gain many other distinctive characteristics called personality traits. Your personality is a combination of inherited traits and acquired traits. Everyone has a personality! Inherited traits are passed on to you genetically by your biological parents. Acquired traits develop from experiences within your environment. **Heredity** and **environment** work together to influence your personality.

Although identical twins share the same heredity, their experiences within their environment may be different.

Heredity and Environment

Heredity
- wavy hair
- smile
- temperament
- body type
- voice
- intelligence

Environment
- mother's prenatal health
- how parents meet physical, emotional, mental and social needs
- school relationships
- peer groups
- community resources
- the media
- religion

The traits that you inherit result from the unique gene set that you received from your parents at conception. You will resemble your family members in many ways; however, you will not have the same set of inherited traits as anyone else unless you are an identical twin. Heredity only determines your potential; how these traits develop is affected by your experiences as you grow and develop.

Religion can provide marker events that signal your passage into maturity within your community.

Your development is influenced by many types of environments. These environments determine the experiences you have and influence your growth and development. When you are young, your environment is provided by your parents. As you get older, you begin to make choices which affect the nature of your environment and how it influences your experiences.

Your family is your first environment but there are many other environmental influences on your development: your school relationships, your community opportunities and resources, the media, information and technology, and religion set standards for living, and provide marker events for many people. Your peer group—those people your own age with whom you share similar interests—might become your most important reference group in adolescence. These and other factors help to shape you into the person you will become.

Understanding your heredity gives you some insight into your potential and helps you identify who you could become. How you choose to develop your inherited traits depends on how you use your environment. Life does not come with a road map, or even with clear directions. There will be many times when you will have to decide which direction you will go. Making good decisions allows you to control the direction that your life will take.

Life is like a game of cards. The hand that is dealt to you represents determinism: the way you play it is free will.

Jawaharlal Nehru

The skills you develop in your teen years can be used to achieve your goals.

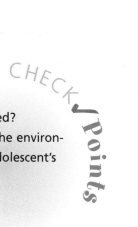

CHECK Points

✓ How are traits inherited?
✓ List the influences in the environment that affect an adolescent's development.

CASE STUDY

Antonietta's Story

At eight months, I was diagnosed with a disease that affects the nerves. As long as I can remember, I have been in a wheelchair.

Every year it seems I have to be hospitalized and I lose something. First, I lost the sight in my left eye as a preschooler, then I lost the strength in my right arm when I was around 12 years old. Another year, I became numb on the right side and was hospitalized at the McMaster Medical Centre for a few weeks. In January, I had another collapse and lost the vision in my other eye. Thank goodness, that did not last and I got my sight back after a month in hospital.

Now I am back at Holy Cross Secondary School, integrated into normal classes. I cope with my difficulties by thinking of good things such as being with my friends and happy people. I am thankful for the things I can do right now that I could not do before. Being able to see everything again is like a miracle. I am really excited about getting an elevator in my house. This will enable me to go up and down stairs and go outside by myself for the first time. The elevator will go directly to the garage and the garage has an automatic door opener. I will be able to become independent like any other teen-ager. When I leave high school, I want to become a secretary or a counsellor.

What do you think?

1. How is Antonietta facing the developmental tasks of adolescence?
2. What are some ways in which Antonietta is demonstrating emotional maturity?
3. Describe how Antonietta's environment will help her in her development.

Reflections and Connections

1. What traits have you inherited from your parents?
2. Explain how you resemble your family. Are the traits you share inherited or acquired?
3. What traits have you acquired as a result of your experiences at school?
4. What opportunities have you used in your community?
5. Describe the first role model you remember from the media.
6. Identify one aspect of technology and explain how it has influenced your development.
7. How has religion influenced your development so far?
8. Who are your peers? What interests do you have in common with them?

Summary

1. Although physical growth ends at adulthood, the co-ordination of skills, known as development, continues through all stages of life.
2. Each stage of life presents challenges for development which are called developmental tasks.
3. Physical development in adolescence includes growth into adult size and form as a man or a woman, and the development of specific physical strengths and skills.
4. Intellectual development enables you to think logically and creatively to solve problems and make decisions.
5. Emotional maturity in adolescence means that you manage your feelings and express them through appropriate behaviour.
6. The ability to co-operate with others in effective relationships is a sign of social maturity.
7. Inherited traits result from genes received from your biological parents, but acquired traits develop due to the influences in your environment.

In this chapter, you have learned to

- outline key aspects of your development in adolescence;
- summarize how your development affects your behaviour;
- explain how your needs change as you develop; and
- explain the influence of heredity and environment on your development.

Activities to Demonstrate Your Learning

1. Draw a line down the middle of a sheet of paper to represent your lifeline so far. List the ages from birth at the top of the line to your present age at the bottom. Identify the marker events and write them beside the age when each occurred. Write negative events to the left of the line, and positive events to the right. Now extend your line another five years. What events would you like to have happen in your future?

2. Create a portfolio of items reflecting your development so far and share it with your teacher and your peers. Try to locate items which reflect your physical, intellectual, emotional, and social development. Include mementos of marker events and influences from your childhood environment.

3. In small groups, design an organizer identifying environmental factors. Suggest how each factor might have influenced the developmental tasks of adolescents in your community.

4. Using social science methods, investigate how your classmates are forming an identity. Formulate a research question. Design and conduct surveys in groups, and compile the results using spreadsheet software. Form a conclusion as an answer to the research question. Publish the results as posters for display in your classroom.

5. Read a biography or novel about an adolescent or young adult. Explain how the individual's experiences are affecting his or her development in each of the four areas. Focus on how the individual meets the challenges presented by the developmental tasks of adolescents.

Enrichment

1. Use on-line searches to develop a bibliography of the books about adolescence available at libraries in your community. Annotate your bibliography with the issues examined in each book.

2. Investigate how development in adolescence is determined by the development of brain cells. Key words to look for: neuron, axon, dendrite, myelin. Draw a diagram to post in your classroom.

3. Recent studies of twins are changing our thinking about inherited and acquired characteristics. Read an article or view a documentary about these studies and report to your class.

4. Interview a teacher to find out how the curriculum you are studying in secondary school is suited to the intellectual development of adolescents. Prepare a list of "study tips" to encourage your classmates to use their schooling to develop intellectual skills.

Here's Looking at You!

Take a look in the mirror. What do you think about your image? How do you want others to see you? Are there aspects of you that cannot be seen on the outside? How you feel about yourself depends on many factors. Imagine that you are much taller, five years older, or perhaps from a different culture. How would that affect your thoughts and feelings about yourself? Would you behave differently? Would you have different friends? Would you feel more or less powerful? Would you feel more "in control" of your life?

By the end of this chapter, you will be able to

❋ outline the relationship between self-concept and behaviour,

❋ describe strategies for developing and maintaining self-esteem;

❋ identify and describe how a personal value system is developed, and

❋ explain the various rights and responsibilities of individuals and families.

You will also use these important terms:

empathy	self-esteem
personal responsibility	temperament
rights	values
self-concept	

Self-Concept

Suppose you were asked to describe the person you look at in the mirror every day. How you define yourself is your *self-concept*.

Your self-concept is your perception of who you are. Your perception is based on attitudes and feelings about yourself. Self-concept can also be called your self-image or identity. All of your life experiences have contributed to your self-concept. Your self-concept includes the inner you—the way you see your-self—and the outer you—the way you choose to present yourself to others.

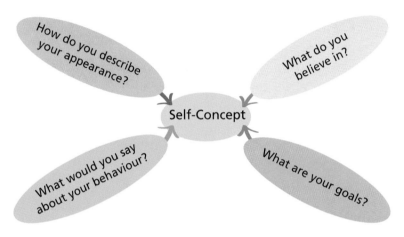

The Inner You

The inner you is the sum of all your thoughts and feelings about yourself. Your inner self sees the sights, sounds, tastes, and smells that you experience every day and determines how you think and feel about yourself in response to them. For example, your height might be 165 cm, but your self-concept might be, "I'm tall," or "I'm short." Your inner you changes as you respond to new experiences. Your inner you colours your perception of yourself and your experiences.

The way you view and interpret a situation is often determined by your temperament. People seem to be born with a tendency to respond in a certain way. Doctors have noticed that newborn babies are not all alike. Some seem happy and optimistic. Others are irritable and cranky. Some are easy going. These differences seem to continue to exist as part of your personality as you mature. Can you identify friends who are cheerful all the time, who always worry about how things could go wrong, or who never seem to worry? This basic quality in people is their temperament. Your temperament affects your point of view, and determines how you perceive yourself and your experiences.

The way you interpret and respond to situations is often determined by your temperament.

Your environment strongly influences your inner self. Your environment includes your family, your culture, and your community. The life experienced by a girl on a family farm in Alberta is very different from that experienced by the daughter of professional engineers from India now living in downtown Halifax. Your families differ, also, in how they accept your behaviour and ideas. If your family were unhappy, they may have been overly critical of you, so that you might have a poor image of yourself. As you mature, other environmental factors such as your schooling, friends, and the media also play a role in how you view yourself.

Temperament reflects the inherited tendencies that you have acquired from your parents. Your temperament determines how you respond to experiences within your environment, both positive and negative. Both temperament and environment influence the perception that you have of yourself. The inner you, the concept that you have of yourself when you look inside, influences the outer you, the image that you create to present to others.

I am not what I think I am.

I am not what you think I am.

I am what I think you think I am.

Cooley's Looking Glass Theory

The Outer You

Your appearance and your behaviour create the first impression others form of you. The outer you is an expression of your self-concept. Your appearance, an important part of the outer you, is determined by many factors. Your body build, height, and colouring are determined by heredity. But how you look is a result of your behaviour. How you style your hair, the clothes you wear, and the way you move are examples of ways in which you can change the outer you. People notice your physical characteristics, your dress, your grooming, and your posture and make judgments about the outer you.

Does the image in the mirror match the image inside you?

What is an image consultant?

An image consultant provides advice to his or her clients in order to help them improve their general appearance, posture, and manner.

What does an image consultant do?

Often image consultants will assist with wardrobe planning to project a successful image for a particular workplace or career. Clients include business professionals, politicians, and private individuals concerned about the external image they present. Image consultants use psychological information, fashion trends, and knowledge of industry factors to make appropriate recommendations to their clients.

Where do image consultants work?

Most image consultants work on a freelance basis and are often hired by individuals and firms on a contract basis. They might also be employed by department stores, boutiques, and other fashion stores.

What does their appearance say about each of these people?

Your behaviour also reflects your self-concept. The activities you choose, how you participate, and how you speak to others are based on how you see yourself. Someone who believes he or she is a good actor might be the first to sign up to audition for the school play and to encourage his or her friends to join also. If he or she does this, others will probably see him or her as an actor. As the old saying goes, actions speak louder than words. Your choices and actions reveal to your family, friends, and others how you view yourself and how you would like others to see you. Your look, your feelings, and your actions add up to your self-concept.

Theodore and Eduardo are best friends. When asked to describe his friend, Theodore says: "Eduardo is tall and has curly black hair. He laughs a lot and likes to tell jokes. We have fun together playing soccer. Ed is a really great kicker, but he never hogs the ball. He loves music and he plays in a band with his cousins. I think that's really cool. Eduardo plays in the school band as well. I don't know where he finds the time! You can always tell when Ed is coming. He wears this cool black hat. It's kind of a trademark."

Self-Concept—The Total You

The way you feel, the way you look, and the way you act are the total you. Your appearance demonstrates choices that you have made and often creates the first impression others have of you. Your behaviour shows the choices that you make in certain situations and, over time, creates a lasting impression. Your appearance and your behaviour combine to reflect your self-concept and determine the image others have of you.

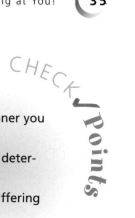

CHECK Points

✓ Describe self-concept.
✓ Distinguish between the inner you and outer you.
✓ Explain how the inner you determines the outer you.
✓ Why do individuals have differing self-concepts?

Reflections and Connections

1. Describe your self-concept, including your appearance and your behaviour, to someone else.
2. Discuss how your temperament influences how you respond to common situations (e.g., it is pouring rain at the end of the school day) compared to how others respond.
3. How has your heredity affected your self-concept?
4. List and prioritize factors in your environment which most influence the inner you.

Self-Concept and Behaviour

Your self-concept influences your behaviour. It serves as a gauge in making judgements of everything you do. How you see yourself in relation to your environment and your experiences creates either a positive or negative feeling about yourself. A positive self-concept can enable you to enjoy your social interactions and feel satisfied with your accomplishments in life. Often, poor behaviour is a result of a negative self-concept. A negative self-concept can make someone difficult to get along with.

Is Your Self-Concept Positive or Negative?

If you have a positive self-concept, you:
- accept who you are and feel good about yourself
- do not worry much about what other people say
- trust your own judgement
- tend to behave in an optimistic manner
- feel confident
- readily take on new challenges
- focus on the positive aspects of any situation
- participate collaboratively in groups

If you have a negative self-concept, you:
- may worry about your shortcomings and failures
- are often afraid to try new things
- tend to have a pessimistic outlook
- feel inadequate and undeserving of positive attention
- are unable to accept compliments
- are sometimes unco-operative
- tend to put yourself down
- often focus on the shortcomings of others

SELF-IMAGE OF TEENAGERS
Percentage Indicating How Well Statements Describe Them in 2000

	Very Well	Fairly Well	Not Very/ Not At All	TOTAL
I am a good person.				
Females	50	46	4	100
Males	53	43	4	100
I have a number of good qualities.				
Females	39	51	10	100
Males	50	43	7	100
I am well liked.				
Females	34	60	6	100
Males	36	56	8	100
I have lots of confidence.				
Females	20	43	37	100
Males	32	47	21	100
I can do most things very well.				
Females	16	61	23	100
Males	29	58	13	100
I am good-looking.				
Females	14	58	28	100
Males	22	57	21	100

Chart from *Canada's Teens* © 2001 Reginald W. Bibby. Reprinted by permission of Stoddart Publishing.

CATHY   by Cathy Guisewite

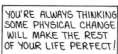

Do You Have a Positive Self-Concept?

Ask yourself: Do I...

- have a lot of good friends?
- like the way I look?
- find it easy to talk to people?
- look forward to the rest of my life?
- usually feel happy?
- have friends that often include me in their plans?
- like to try new things?
- see the good in people?
- laugh at my mistakes when I fail?

Developing a positive self-concept begins in childhood. As children learn new things, they feel good about their accomplishments. Suppose a child is learning how to do up the buttons on her sweater. With each try, she is praised by family members and encouraged to keep trying. When she is finally successful at doing up all the buttons properly, she will be complimented about her skills. She will feel proud of her accomplishment. Experiences in childhood provide the foundation for a positive self-concept.

Self-Esteem—If You Build It, Success Will Follow!

The way you define yourself is your self-concept. It is actually the picture or image that you see in your mind's eye when you imagine yourself. How you feel about that image is your self-esteem. Your self-concept shows the image while your **self-esteem** judges the value of that image. It is certainly tied to your self-concept, but it is even more. It is based on how you are treated by others and how you treat yourself. Self-esteem is developed over time. Your self-esteem can be high or low depending on how valuable you judge yourself to be. If those around you accept and praise your efforts, both your self-concept and self-esteem will be affected positively.

Self-esteem is the tool that can mold your success. People with high self-esteem believe themselves to be valuable. Therefore, they have more confidence and challenge themselves to get the most out of life. Having high self-esteem does not mean being conceited, however. If necessary, you can take action to improve your self-esteem. Improving your self-esteem will enable you to face challenges and experience success in life.

Where do your feelings of self-worth come from?

Successful experiences that are recognized by others help to instill a positive self-concept.

Take responsibility
- Learn to deal with the many demands of your daily life. Managing these demands successfully will boost your self-esteem.

Learn from your experiences.
- View mistakes as part of the learning process. Use them to figure out how you might do things differently in the future.

Accept praise.
- When someone pays you a compliment, allow yourself to feel good. Usually praise is well deserved.

Use your strengths to support others.
- Co-operating with others often provides a sense of accomplishment and helps you feel good about yourself.

Focus on your strengths.
- You know that there are certain things that come easily to you and that you can do well. Identify your skills and talents.

Accept yourself.
- We are all unique and no one is perfect. You accept your friends as they are–why not do the same for yourself?

To love oneself is the beginning of a lifelong romance.

Oscar Wilde

Self-confidence is the first requisite to great undertakings.

Samuel Johnson

Reflections and Connections

1. Think of someone you know who has a positive self-concept. How does that person interact with others?
2. How does another person's self-concept affect you?
3. What are some ways in which self-concept can affect behaviour?
4. Explain how a positive self-concept is formed.
5. Why do some people have poor self-esteem? What could you do to boost someone's self-esteem?

CHECK Points

✓ Describe the relationship between self-concept and self-esteem.

✓ What are the characteristics of self-esteem?

✓ How might you recognize low self-esteem?

✓ What steps can you take to improve self-esteem?

You and Your Values

Your self-concept and your self-esteem are greatly influenced by what you believe in. The ideas and beliefs that guide your life are your *values*. When you value something, such as honesty, your behaviour shows it. For example, if you are given too much change in a store, you bring it to the attention of the clerk and return the money. Your values are part of the inner you, but they are reflected in all your actions and decisions.

Most values fall into two main groups. Tangible values are those things that you can see and touch. Tangible values might include clothes, money, and cars. Intangible values, such as love and honesty, cannot be touched or seen. You might think that you can see love, but what you are seeing is the behaviour that results from valuing love, such as a kind act or a warm embrace. Understanding the difference between tangible and intangible values is not often easy.

Tangible and Intangible Values

Tangible values	Intangible values
• house, car, pets, jewellery, books, money, stereo, clothes, video games, artwork	• trust, independence, friendship, education, love, religion, creativity, knowledge

Some values simply affect our personal preferences. For example, Nicki and Amita both wanted to be involved in school activities. Nicki values leadership so she joined the student council. Amita, however, values creativity, so she joined the art club. Values helped both girls make their choices. Were either of them right or wrong? better or worse? With these kinds of values, there is no right or wrong. They are both acceptable for making choices.

Many values are viewed as either right or wrong. Most often these are the intangible values, such as honesty, dependability, trust. For example, love and patience are seen as positive values, whereas dishonesty and greed are seen as negative values. The interesting thing about values is that they are not always clearly positive or negative. Debating issues can help you to clarify your values.

The choices that you make reflect your personal values.

Complex values

Some values are not always clear. For example, you may believe that killing is wrong, but what about killing in self-defense? Or you believe strongly that people should not steal, but what if someone steals food for starving family members? When there are many human issues involved, values may be in conflict. Careful consideration of what is truly important to you will help to clarify your values.

Learning Values

Your family, your friends, and your community all play a part in the development of your values. You began learning values from the moment you were born. During adolescence you are clarifying your values. You now have the opportunity to examine the influences of your family, your community, and your peer group in order to develop a personal value system.

Your family provided the foundations for many of your values. For example, if parents value independence, they may focus on teaching children to do things for themselves at a very early age. If they value creativity, they may

focus on helping their children to draw, paint, dance, and sing. All of the behaviours of your family members were based on their values.

Your community can influence your values. For some people, religious beliefs and practices provide guiding values. Your experiences at school may affect how you value education as well as your skills. Watching a specific movie, news item, or television commercial may influence what you believe to be important. Interaction with people who have strong cultural values, media that tries to influence what is important to you, and organizations which are seeking your participation all have an impact on what you choose to value.

Values are demonstrated by your peers' actions. Their behaviours may show that they value freedom, courage, or fairness. Many friends in your peer group may share your values; however, they may also hold differing values. These situations can cause conflict in groups. When your values are different than those of your friends, you might need to make a change. You might need to choose between changing your values, or finding new friends.

Families provide activities that reflect their values.

CASE STUDY

Alana's Story

Alana had made a new friend in her family studies class. Minh was really fun to be with. They shared many interests and liked the same fashions and TV shows. Frequently, Minh would meet a group of her friends at the food court in the mall. Alana was invited to join Minh and her friends.

Alana had a really good time with the group, but did not like the fact that they all smoked. Minh and her friends were constantly offering Alana cigarettes.

Alana, who had grown up in a smoke-free home, felt the pressure to smoke in order to be part of the group. She gave the decision a lot of thought. The next time Alana was with the group at the mall, she let them know that she did not mind if they smoked, but she was not a smoker and was not going to start. Some of the girls in the group teased her, but soon, it was no longer an issue. No one offered Alana cigarettes anymore and accepted that she was not a smoker.

What do you think?

1. What values did Alana take into account when she made her choice?
2. Do you think Alana's family had any influence on her values?
3. What risks did Alana take in making her choice?

Your Value System

During your adolescence you are developing a personal value system. You may feel pulled in many different directions by conflicting values. You will be faced with making decisions based on your values on a daily basis. There are many perspectives to consider when choosing your personal values:

- Is the value positive or negative? Will it bring harm or pain to me or to others? Always aim for the positive!
- Is the value something that is deeply rooted in my family? If so, carefully consider the importance of the value to you as a family member. Sometimes values shared by a family create a strong positive bond.
- Where does the value come from? Is it based on the common good or is it based on selfish motives?
- Is the value moral? Is your conscience telling you that the value is right?

Morals
what are they?

Morals are principles of right and wrong that you live by. A strong sense of values helps you to choose the principles with which you will make moral decisions. Principles are higher values that summarize your beliefs of what is right and what is wrong for people in society. The values and principles that you hold make up your moral code.

Your value system is an important part of you. It affects who you are, but it also affects others. Once you have a strong value system in place, taking action will be easier because you will know what you believe in. Your family and friends are affected every day by the choices and behaviours resulting from your personal value system.

CHECK Points

- ✓ What are values?
- ✓ Distinguish between tangible and intangible values.
- ✓ How do your values affect your life?
- ✓ How do people acquire their values?
- ✓ Why do people sometimes have conflicting values?

The decisions that your friends and family members make are also based on their value systems. Some of their value systems may differ from yours. These differences are sometimes difficult to understand. *Empathy* is understanding someone else's feelings or point of view. By practising empathy, you put yourself in the other person's shoes. Understanding the values of others can help to foster important universal values such as respect, tolerance, and compassion.

Conduct Social Science Research

What do your peers value? In order to understand the values that your peers hold, do some primary research. Design a questionnaire using information in this section. In your research you may include questions such as:

1. **What is the most important belief or value that you hold?**

2. **What are the most important values held by your age group today?**

3. **How does your behaviour show your values?**

4. **What would you do if you won a million dollars?**

5. **What is more important, the economy or the environment?**

Reflections and Connections

1. What values have you learned from your family? How did you learn them?
2. What values are emphasized in the curriculum at your school?
3. Think of a recent startling news item. Did it affect your values in any way? If so, how?
4. What values do you share with your peers? How do these values affect how you behave with your friends? Do some of your peers hold values that you find questionable?
5. Which do you believe to be more important, friendship or honesty? Why?
6. Describe a situation where it was important for you to understand someone else's values.

Personal Rights and Responsibilities

As an adolescent forming an identity, you are less dependent on others and are making more decisions about your own life. As you mature, you are also required to be responsible for your choices. **_Personal responsibility_** means that you are reliable and accountable for the decisions that you make. This means that you are willing to accept the consequences of your behaviour. If you are accountable, you will acknowledge your mistakes and not blame others when things go wrong. People who are responsible and accountable use phrases such as, "I made a mistake, but I will fix it as soon as I can." Others will trust you if you demonstrate personal responsibility by meeting your commitments.

Unlike responsibilities, the rights we have as individuals change very little as we age. We all are entitled to rights, no matter what our age. **_Rights_** are usually defined by laws and charters but must be defended and promoted in our daily lives. As children, we are not able to defend our rights. Children must rely on adults to ensure their rights are being maintained. As we grow older, we are more able to exercise and uphold these rights for ourselves. As an adolescent, we become more responsible for knowing and upholding our rights.

Above: As a child your parents selected your clothing. You had little to say about what clothes were bought for you or what you wore on any given day.

As an adolescent you can choose what you wear each day. Often you have input into what type and style of clothing is purchased for you.

	Childhood	Adolescence	Adulthood
Clothing	dressed by parents	select own clothes	purchase and select clothes for self and family
Education	learning required and adult-directed	learning required, but self-directed	learning is a self-directed choice
Money	no income	limited income	self-determined income
	little, if any, spending choices	some spending choices	range of spending choices

Changing Choices—Changing Responsibilities

When you volunteer to take on a household task such as doing the laundry, you are demonstrating personal responsibility within the family.

Responsibilities

There are many ways that you can demonstrate responsibility. Can you describe some specific things that you can do in these situations?

- **Share household duties.**
- **Complete school work.**
- **Being environmentally accountable.**
- **Controlling household spending.**

To be nobody but yourself, in a world which is doing its best, night and day, to make you into somebody else, means to fight the hardest battle which any human being can fight, and never stop fighting.

e. e. cummings

What is the Canadian Charter of Rights and Freedoms?

The Canadian Charter of Rights and Freedoms is one part of the Canadian Constitution. The Constitution is a set of laws containing the basic rules about how our country operates. For example, it contains the powers of the federal government and those of the provincial governments in Canada.

The Charter sets out those rights and freedoms that Canadians believe are necessary in a free and democratic society. Some of the rights and freedoms contained in the Charter are:

- freedom of expression the right to a democratic government
- the right to live and to seek employment anywhere in Canada
- legal rights of persons accused of crimes
- Aboriginal peoples' rights
- the right to equality, including the equality of men and women
- the right to use either of Canada's official languages
- the right of French and English linguistic minorities to an education in their language
- the protection of Canada's multicultural heritage.

Human Rights Directorate of the Department of Canadian Heritage

Go back to the mirror. Take another look. What you see is the outer you, a combination of your heredity and what you want others to see. Although your appearance reflects your self-concept and values, your behaviour is the strongest indicator of your values. As you progress through adolescence, it is important to remember "it's what's inside that counts!"

CHECK **Points**

- ☑ What is the definition of responsibility?
- ☑ How are rights and responsibilities related?

Reflections and Connections

1. How have you been able to uphold your rights recently?
2. List some responsibilities that you have now that you did not have as a child.
3. How does your personality affect the way that you act on your rights and responsibilities?

Summary

1. Your behaviour is the external demonstration of your self-concept.
2. You develop high self-esteem through the reflections of your family and friends when you focus on your strengths and take responsibility for your actions.
3. Your personal value system is based on choices you make as a result of your environment and your experiences.
4. Although basic human rights remain constant throughout life, responsibilities and the right to make certain choices change as you get older.

In this chapter, you have learned to

- outline the relationship between self-concept and behaviour;
- describe strategies for developing and maintaining self-esteem;
- explain the various rights and responsibilities of individuals and families; and
- identify and describe how a personal value system is developed.

Activities to Demonstrate Your Learning

1. Select a character from a current television drama. Describe the character's traits, their self-concept, self-esteem, and values. Analyse how these various characteristics affect and direct the character's behaviour.
2. Read an article that describes ways in which parents help their children to build self-esteem. How are their methods similar to or different from the ways that adolescents can elevate their self-esteem?
3. Use social science research methods to conduct a survey of your peers to identify the most important factors that have influenced their personal values. Report the conclusions in a poster.
4. Locate sources of statistical information such as Statistics Canada or other published surveys. What do adolescents value across the country? Do their values differ from those of your peers? What values are common for your age group? Chart the results of your investigation.

Enrichment

1. Read a biography of a famous folk hero. *Robin Hood* Explain how the character's self-concept directed her or his actions and behaviour.
2. Survey a group of seniors to find out what their values are. Compare their answers to those you received from your peers. Which values do both groups share? Why do you think differences exist? Create a chart to show the similarities and differences of the two groups.
3. Go to a library or bookstore. Survey the self-help section to find titles that help to elevate self-esteem. List some unique strategies that are suggested to boost self-esteem. Analyse whether or not these strategies would be useful to raise your own self-esteem.

Where Are You Going?

Where are you going? What do you want to do this week? this year? in your lifetime? Do you dream of being an NHL star? Do you dream of discovering a cure for cancer, climbing Mt. Everest, or running a marathon? Perhaps your dreams include graduating from college, marrying a wonderful person, and having children. You might be dreaming right now about mastering polynomials and finally cleaning your room! Achieving your dreams, at least some of them, requires setting goals and making some choices.

By the end of this chapter, you will be able to

* identify your personal short-term and long-term goals;

* explain the importance of making choices in order to achieve your goals;

* describe strategies for making informed and responsible decisions; and

* apply decision-making models to making choices in your life.

You will also use these important terms:

alternatives	needs
choices	resources
consequences	standards
decision making	wants
goal	

Developing Decision-Making Skills

Identifying Your Goals

What are you going to do after school today? What are your plans this weekend? What mark do you want on your next science test? What do you want to accomplish this year? How do you hope to spend your life? The answers to these questions are some of your goals. A *goal* is something you want to achieve at a certain time in your life. Your goals are based on your needs and your wants. Setting effective goals enables you to make choices that will help you to meet your needs and to get the things you want in life.

Needs and Wants

 Needs are those things that are necessary for your growth and development. You have needs in all areas of your development. Some of your needs are more important now than others and your needs will change as you get older to meet the challenges of adulthood.

Wants are those things that will make your life more pleasant. Your wants will differ from those of other people. Your friends may think your wants are crazy! The things that you want reflect your interests and values and form part of your self- image. Although you can live without them, having the things you want improves the quality of your life.

Needs and Wants

Needs	Wants
• food	• purple T-shirt
• exercise	• canoe down the Winisk River
• shelter	• eat the hottest chili
• safety	• learn to play the bagpipes
• family	
• friends	
• a sense of accomplishment	

Short- and Long-Term Goals

If you hope to achieve something in the near future, it is a short-term goal. Short-term goals can be accomplished in a short period of time. Long-term goals will take longer to achieve. Long-term goals might take years to accomplish.

It is easier to achieve long-term goals if they are broken down into several short-term goals that will take less time to achieve. For example, Cadina may begin by playing hockey on

Self-Fulfillment

Esteem

Love and Belonging

Safety and Security

Survival

Maslow's Hierarchy of Needs

Short- and Long-Term Goals

Short-Term Goals

- Sarah hopes to get her hair cut after school today.
- Mohamed would like to finish his Geography project this weekend.
- Anna wants to save the money to buy a concert ticket so that she can go with her friends.
- Matt wants to find a part-time job this month.

Long-Term Goals

- Andrew wants to graduate from high school and enter university to study law.
- Priya hopes to finish school and join her father in the family business.
- Chris looks forward to owning his own car.
- Cadina dreams of playing for the Canadian women's hockey team in the next Olympic Games.

the school team and in a community league for a few years to develop her skills, and then select a university with a strong women's hockey program. Each of these smaller, short-term goals is a step toward her long-term goal of playing on the Canadian women's Olympic team.

Setting long-term goals and short-term goals enables you to develop a plan of action for using your resources to achieve what you want in life. Goals enable you to meet your needs and to get the things you want. Establishing goals is an important part of figuring out who you are and the roles you will play in life. Worthwhile goals are based on the values and beliefs that are important to you. These values determine the **standards** that you will use to determine whether or not you have achieved your goals.

> *Goals are dreams with a deadline.*
>
> Gail Vaz-Oxlade

SMART Goals

- **S**pecific
- **M**easurable
- **A**ttainable
- **R**ealistic
- **T**imely

Adapted from *The Learner's Edge*

Developing your skills in a league can be a step towards playing on the Canadian Olympic team.

Reflections and Connections

1. Make a list of needs in each of the areas of your development at this stage of your life.
2. List ten things you want in your life now. Compare your top ten list with those of your classmates.
3. Select one of the needs of adolescents, and identify "SMART" long-term and short-term goals that meet that need.
4. Ask your parents to list ten things they want in their lives. Compare their lists with yours.

CHECK Points

✓ What is the difference between needs and wants?

✓ How do short-term goals relate to long-term goals?

✓ Explain how your self-concept affects your goals.

Getting to Your Goals

Each step along the way to achieving your goals involves making *choices*. Selecting an option when there are many alternatives available to you can affect your present and future quality of life. Setting your short- and long-term goals requires you to make selections on a regular basis. These choices are based on a sound knowledge of yourself as well as a thorough understanding of the decision-making process.

VALUED GOALS OF TEENAGERS AND ADULTS
Percentage Viewing as "Very Important" in 2000

	Nationally	Males	Females
Friendship	85	80	90
Freedom	85	84	85
Being loved	77	65	87
Having choices	76	73	79
A comfortable life	73	71	74
Success in what you do	71	70	73
Concern for others	62	51	73
Family life	59	51	66
Excitement	57	58	55
What your parents think of you	44	38	50
Your looks	39	40	39
Recognition	32	34	31
Spirituality	29	23	35
Having power	24	32	16
Being popular	16	21	11
Religious group involvement	10	9	10

From *Canada's Teens* copyright 2001 Reginald W. Bibby. Reprinted by permission of Stoddart Publishing.

As part of your transition towards independence, you will have the freedom as well as the responsibility of making more and more decisions. These decisions will affect your daily life as well as your future. It is important to acquire skills in the decision-making process as an adolescent. Mastering decision-making skills will help you to become more independent. Good decision making will also allow you to have control over many areas of your life.

Decision Making Is a Process

You make decisions on a daily basis. Decisions impact all aspects of your life—money, time, friends, work, health, and family. ***Decision making*** involves choosing between options. Some decisions are major. They have important and often long-term effects on your life. An example of a major decision would be dropping out of school. Other decisions are minor and do not have a great impact on you in the long term. An example of a minor decision would be going to a movie rather than to a concert.

Mastering decision making allows you to take control of many areas of your life

CASE STUDY

Silvana's Story

Silvana knew that getting good grades was important to getting a good summer job with the local community centre's enrichment program. She needed the job because the money would be really good and the reference would be important for her resume. She had been working hard at the beginning of the term, but recently her marks in Math had not been good. Silvana realized that she was spending so much time playing computer games that she was rushing her homework.

The term was more than half over and Silvana's goal of getting good grades was beginning to look out of reach. Silvana thought about asking her parents to hire a tutor to help her catch up. The extra help might allow her to continue spending time on computer games. She also considered giving up playing video games and spending more time working on her Math homework. Silvana's friend Josie suggested that she could just drop Math, so that the low grade would not be an issue this term at all.

Silvana can make decisions about her situation in many different ways. Many factors will come into play as she makes her choices.

What do you think?

1. How might you choose among these options?
2. Are there other possible options available in this situation?

Reflections and Connections

1. List some of the decisions that you now make as an adolescent that you did not make as a child.
2. Give examples of some major decisions. How could they affect your life in the long term?
3. Describe some minor decisions you make every day. Why do they not have a great impact on your future life?

How do I make decisions?

You make many decisions every day. The way you make each decision may be different.

Some Ways that Decisions Are Made

1	**Habit**	We all have things that we do in personal and unique ways on a regular basis. It is the way you usually do things. You take the same route to your after-school job every day. Sometimes these habitual decisions provide you with shortcuts. You do not have to spend time thinking about making a choice. Sometimes these decisions can just get you into a rut.
2	**Custom**	Following what others usually do in a given situation is often a result of custom. These decisions are often affected by culture and heritage. Choosing to participate in a religious ritual or wearing a particular clothing symbol are examples of customs.
3	**Imitation**	You make the choice that is frequently made by others. This can be a good way to make a decision, such as when you decide to join many of your peers in a study group after school. However, you must be careful to imitate only if these choices are good for you. Decisions made by others may not always suit your needs, lifestyle, or talents.
4	**Impulse**	You are in a mall and see that your favourite musical group has just released a new CD. You buy it without a thought. In this case, you are not allowing yourself any options but to do what you feel like in the moment. If you make major decisions on impulse, you may not be allowing yourself enough time to consider your options and choose the best course of action.
5	**Coin toss**	The burger or the chicken fajita? This could be decided by a coin toss, leaving the outcome to chance. Leaving choices to chance may be a good strategy when the outcome has no real or long-term implications.
6	**Default**	This is choosing not to make a decision. An example might be not selecting a particular topic for a class presentation, but merely taking the only one left after everyone else has chosen one. Sometimes default decisions are made due to procrastination.

Many decisions can be made lightly, especially when there are no long-term effects. Methods such as the coin toss can be used to make the decision about which movie to see. Many choices and decisions that you make on a daily basis can be made by these methods. Decisions such as which sneakers you will wear, which cereal you choose to eat for breakfast, or which comedy show you choose to watch on TV, do not affect others and have little, if any, affect on your life.

There are decisions made on a daily basis that affect you and others around you. These decisions require more than just a coin toss or an impulse reaction. They require you to act in a way that reflects your values, your inner self, and the way you want to be perceived by others. Examples of these types of decisions might be: which extracurricular activity you will participate in, which university you will go to, or whether or not you take on a part-time job. A way of making decisions that provides for the best possible end result is to employ a step-by-step process that involves lots of input and consideration and a minimal amount of risk.

Reflections and Connections

1. Compare with a friend the types of decisions you make out of habit on a regular basis.
2. Describe a recent decision you made on impulse and the effect it had on you and others. Why do you think you make these decisions habitually?
3. Think about a decision you made by imitating others. Did you find the result of the decision personally satisfying?

The Decision-Making Process

The decision-making process is a step-by-step method you can use to guide your thinking. It helps you to take into account all the important aspects of your decision. The process ensures that you make the best possible decision with the information that you have. Each step in the method helps to organize your information as well as your thought process.

Nothing is more difficult, and therefore more precious, than to be able to decide.

Napoleon Bonaparte

The Decision-Making Process

1. **Identify the problem.**

   Why is there a need for a decision? What is the problem to be solved? Is there a choice to be made? It is important to clearly define the situation that requires a decision. If you state your problem as a goal, it may force you to deal with conflicting values.

2. **Consider what the standards need to be for a successful result.**

   By being clear about the standards that you expect for your goals, you are defining how you will measure success, and making the decision-making process clearer.

3. **Identify all of the possible alternatives.**

   Make a list of all of the various *alternatives* that may be available to you. This is the time to be creative and let your ideas flow freely. This is also the time to list all of the *resources* you have to help you make your decision or solve your problem.

4. **Predict the consequences of each alternative.**

   Each option that you have identified may have some positive as well as negative *consequences*. You must assess the positive and negative consequences of each of the alternatives. Then you can determine which of your possible alternatives has the most positive consequences.

5. **Estimate the probability of each alternative and the related consequences.**

   Considering the circumstances and the resources you have, how likely is it that the predicted consequences will happen?

6. **Choose the best alternative and take responsibility for your decision.**

   This is the point at which your values, goals, and standards are really put into action. The alternative you choose may not be the easiest or most popular, but to be effective, it should meet your needs according to your values .

7. **Carry out the decision with a plan of action.**

   There may be many things to do in order to implement your decision. It is a good idea to develop your plan and imagine each step. By doing this, you may be able to identify any obstacles that you encounter. With this type of forward thinking, you have ways of overcoming obstacles before they happen.

8. **After you have implemented your decision, you need to evaluate the results of your actions.**

   Evaluate how well the decision turned out and you will increase your skill in making future decisions. Did you explore every possible alternative? Did you have enough information about each alternative? Did you effectively follow your plan of action? Did the results of your decision meet with your personal values and standards?

career Link

What is a manager?

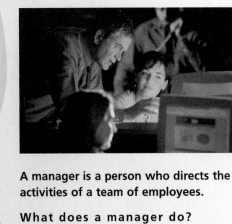

A manager is a person who directs the activities of a team of employees.

What does a manager do?

A manager spends the day making decisions and implementing plans. The kinds of decisions that are made depends on the type of company the manager works for and the department that the manager is in. For example, a human resources manager would decide on who to hire for a particular job within a company, make decisions about the best benefits plans for the company employees, and may be responsible for planning and implementing safety practices and security policies. These managers are also responsible for decisions regarding grievances, layoffs, and firings. Through their decisions, human resources managers are responsible for maintaining the standards of the company that they work for.

Where do managers work?

Managers work in virtually every type of business, industry, or community organization. Most large companies and organizations have human resources managers. Regardless of the title, any business or organization that employs people to make decisions and implement plans for that company is requiring that person to be a manager.

Once you have clearly defined the need for a decision, the path of the decision-making process becomes clearer. Setting a goal of improving your grades by ten percent will give a clear direction in exploring your alternatives.

Standards
what are they?

Standards define what we deem to be acceptable and unacceptable. For example, you have a certain standard of cleanliness in your personal environment. You may not mind if things are a bit disorganized in your room, but you draw the line at big "dust bunnies." Individuals all have different personal standards. Standards are very closely linked to values.

Exploring all of the alternatives gives you a better chance of making the correct choice. When you take your values, goals, and standards into account, you usually are able to make a good decision.

CHECK✓Points

1 ✓ Why is it important to clearly identify the problem or issue before you look for alternatives?

2 ✓ When exploring alternatives, why would you include ideas that are unconventional?

3 ✓ Explain why it is important to consider the consequences of each alternative.

4 ✓ When you evaluate the results of a decision, how do you know that it has been successful?

The decision-making process is an important tool for implementing your short- and long-term goals. Look at the case study below for an example of how the process can be used.

CASE STUDY

Omari's Story

Omari is thrilled. He just found out that he made the basketball team at school. After the joy of his success, he realizes that being on the team will mean lots of after-school practices and games. He also takes part in other activities. He attends Cadets every Tuesday night and that commitment also includes occasional weekend activities. He is part of the school swim team. Since he has been on the team for two years, many of the younger swim- *mers rely on him for help and guidance during their practices two nights a week. Omari has been able to maintain an average grade of 80 percent in his school work, but this success has come from lots of effort. He usually spends about two hours on homework every night and at least four or five hours on the weekend. He now identifies a problem by asking himself, "When am I going to have time to do my homework?"*

What do you think?

1. What are the issues facing Omari in his decision-making process?
2. What are some of the standards that Omari may want to maintain?
3. Describe the decision-making process for several alternatives that you have identified for Omari.
4. For each of the alternatives, identify how the success of the decision might be evaluated.

Gino is very pleased with the results of his plan of action. Deciding to eliminate video games and game shows from his daily schedule has given him much more time to spend with his friends.

Reflections and Connections

1. Describe a good decision that you made recently. Why do you think it turned out well?

2. Think about a decision that did not turn out well for you. Which of the steps in the decision-making process were problematic?

3. Identify ways in which you have taken responsibility for your decisions.

PEANUTS

DO YOU EVER THINK MUCH ABOUT THE FUTURE, LINUS?

OH, YES... ALL THE TIME

WHAT DO YOU THINK YOU'D LIKE TO BE WHEN YOU GROW UP?

OUTRAGEOUSLY HAPPY!

Scare Tactics Won't Stop Teen Smokers/Kids Who Light Up Need Some Sort of Immediate Penalty, Not Vague Predictions

By Rachel Sa

As a rule, teenagers are not stupid. No really, we're not.

Sure, a lot of teenagers smoke and, granted, smoking is nothing if not the very definition of stupidity. It will kill you. We all know this, we've all accepted it (well, those who haven't are having denial issues). Smoking causes lung cancer, heart disease and emphysema, just to name some of the really nasty ones that are printed in huge bold letters across each and every package.

Smoking also shrivels your skin, yellows your teeth and stains your fingers. Oh, and let's not forget that delightful odour that clings to your clothes (yummy).

Toss into the fold that in the past few years smokers in our society have become public enemy No. 1—ostracized and vilified, banned from restaurants and malls and mostly all public places and generally frowned upon by the righteous non-smoking community. And yet many teenagers are still making the decision to start smoking.

Okay, so maybe some teenagers are stupid—or at least, they make some stupid decisions. But I still don't think it's an uninformed decision.

What bugs me most about the countless "Stop Smoking" campaigns making the rounds in print and on television is that they assume that kids smoke because they don't know the dangers. I'm pretty sure the teens who smoke know all about the dangers—it's a rare thing to have a school without an anti-smoking campaign of some kind.

They just don't care about the dangers. (Cancer? Heart disease? That sucks—but it won't happen to me!) Remember? We young people have this "I'm going to live forever, nothing can hurt me, I am invincible," kind of mentality. It's the same reason why

some kids make the even more idiotic—and deadly—decision to drink and drive. (Sure, smoking kills—other people. Not me though. Puff, puff.)

And now we're getting visual aids. Goody! Pictures to go along with the really big warnings—another thing to ignore. Okay, I'm sure these new photos of cancerous lungs and gums will definitely have an "ick" factor. But hey—they're not "our" lungs. Besides, these won't be the first graphic, vile and disgusting images young people have ever been presented with. Movies, TV and the Internet have all taken care of that.

So I think it's about time the anti-smoking crusaders realized that scaring kids off cigarettes with predictions of a horrible death in the distant future won't have the desired effect.

Strengths and Weaknesses

Stop playing to our strengths—denial and indifference. Try playing to our weaknesses—our craving for acceptance and our fear of rejection. Sad as this may sound, I truly think convincing kids that no one will sit with them at lunch if they smoke, would be a lot more effective than the death threats.

Kids who smoke need some sort of immediate penalty, not vague predictions of disease years down the road. How about: "Warning—Smoking will make you impotent—who will want you then, stud?" Or, "Smoking will make those cute boys over there point and laugh at you."

Let us not underestimate the power of body image. Puberty and adolescence are the times when young men and women are feeling most self-conscious. The anti-smoking crusade should be playing to those weaknesses. How about, "Smoking may give you explosive acne?"

Studies have already suggested the most effective of the new pictures on packs of cigarettes is the photo of the diseased gums. No one wants to have an actual, visible flaw.

Everyone wants to be one of those beautiful, slender women in the ads, delicately holding their slender cigarettes while beaming at the camera with pearly white teeth. Air brushed to perfection. If those ads ever pictured an actual smoker—the creased, yellowed skin, the stained teeth and dry hair, it would be that much harder for young girls to delude themselves into thinking cigarettes would not hurt them.

Some television shows have already started hopping on this bandwagon. It is rare that you will find the good guys lighting up on TV anymore. That loathsome habit is left to the villains now. Although even that does not always work, because Buffy the Vampire Slayer's arch nemesis, Spike, still looks pretty cool when he lights up. But the media would do well to focus attention on that sort of campaign. Maybe talk with the geniuses who came up with those catchy GAP commercials. That would really get our attention.

All that said, one caveat: What sort of society are we becoming when we have to teach our young people to value their looks over their lives? Not a very healthy one.

Toronto Sun, 01-24-2000
Reprinted with the permission of Sun Media Corp.

Summary

1. Setting goals enables you to plan to meet your needs and get the things you want in life.
2. Short-term goals form steps along the way to achieving long-term goals.
3. You will have to make good decisions in order to use your resources to achieve your goals.
4. You choose from many ways of making decisions as you make choices in your day-to-day living.
5. Important decisions should be made step by step considering your standards for a good solution and the consequences of all the available alternatives.

In this chapter, you have learned to

- identify your personal short-term and long-term goals;
- explain the importance of making choices in order to achieve your goals;
- describe strategies for making informed and responsible decisions; and
- apply decision-making models to making choices in your life.

Activities to Demonstrate Your Learning

1. Develop a plan for achieving a goal that you have set for yourself this year.
2. Use social science research methods to determine the most important long-term goals for students in your community. For the top five, find out what short-term goals individuals have set for achieving their goals.
3. Invite a panel of senior students to speak to your class. Ask them to discuss the factors to consider when making important decisions in you next few years of secondary school.
4. Interview a successful person in your community. Find out about how they implemented their goals. Describe the strategies that they used in the decision-making process.

Enrichment

1. Conduct a survey to determine which of the decision-making strategies is used most often by various age groups. Graph the results using spreadsheet software and present your conclusions.
2. Interview a senior citizen in your family or community. Find out about some important decisions they have made in their lives. Are they happy with the decisions that they made? Why or why not?
3. Read a biography of a famous person. Summarize some of the major decisions that they made in their life. Report your findings to the class.

CHAPTER

5

Relationships in Your Life

You will meet many people on your journey through life. How will you relate to them? Do you need them to meet your needs? Will you form friendships that last a lifetime or will you spend time with them for a while, then go off on separate paths? How can you improve your relationships with people, whether you want to become friends or are just assigned to work together? How do you recognize real friends, and what do you do when you do not get along? What are the secrets of positive relationships?

By the end of this chapter, you will be able to

✳ explain the importance of relationships for meeting social and emotional needs;

✳ describe the characteristics of healthy and effective relationships;

✳ explain how a variety of relationships meet different needs; and

✳ explain why people need to belong to and participate in groups.

You will also use these important terms:

affection	rapport
commitment	reciprocity
compromise	relationships
functional relationships	respect
peers	roles

You Need Relationships

How important are the people in your life? If you are like 85 percent of Canadian teens surveyed by Reginald Bibby in 2001, you probably consider your relationships to be very important in your life. **Relationships** are the bonds formed between people based on common interests, and often, on **affection.** People with whom you have relationships share your experiences, both your pleasures and your pain. Relationships are based on **reciprocity**—giving and receiving. You cannot really have a relationship with someone who does not have a relationship with you. Nor can you have an effective relationship in which one person is doing all the giving, and the other person is doing all the receiving. The people you form relationships with are your reference group. They provide you with feedback on who you are. Your relationships can support your self-concept and provide you with companionship.

IMPORTANCE OF FAMILY IN 2001

	Nationally	Males	Females
VERY IMPORTANT			
Friendship	85	80	90
Freedom	85	84	85
Family Life	59	51	66
What your parents think of you	44	38	50
HIGH LEVEL OF ENJOYMENT			
Friends	94	93	95
Your mother	71	65	76
Your father	62	58	66
Brother(s) or sister(s)	58	53	61
Your grandparents	54	50	58

Chart from *Canada's Teens* © 2001 Reginald W. Bibby. Reprinted by permission of Stoddart Publishing.

Why are relationships so important in your life? Friendships are a major source of enjoyment for teenagers. Meeting new people allows you to develop your interests and share experiences with others. Discussing your opinions, thoughts, and feelings with your friends helps you to clarify your ideas. Friends may encourage you to try new activities and argue other points of view to challenge your thinking. A supportive friend might prod you to reconsider your options before you act. Relationships can provide you with good times and enrich your life.

Belonging to a group of friends meets your need to be accepted by others. You will develop a greater understanding of your own emotions and develop empathy by sharing the feelings of others. You will learn to be sensitive to the thoughts and feelings of other people. Because friends are concerned about each other's welfare, you will feel cared for and secure. In return, you will contribute to your friends' sense of love and security. Mutual acceptance and mutual support enable you to spend time co-operatively with others in the group.

CHECK Points

1. What are the characteristics of relationships?
2. Explain why relationships are necessary to enable you to develop socially.
3. How do relationships meet a person's emotional needs?
4. Explain how relationships can enrich your life.

Reflections and Connections

1. Think about a friend that you have now. What activities do you enjoy doing together? Do boys and girls enjoy the same types of activities with their friends?

2. Describe a situation when a friend encouraged you to do something. Why did you need the encouragement? Did the situation turn out well for you?

3. Has a friend ever encouraged you to reconsider you actions? Why? Did your friend's advice improve the situation for you?

Qualities of Relationships

Effective relationships begin with **rapport**, the feeling of being comfortable in each other's company. You feel comfortable with people who accept you for who you are, who tolerate your ideas and interests, and who can be happy for you. Good friends show their **respect** for you by expressing their own point of view even though it may be different from yours. Often, rapport is based on affection. This comfortable give and take supports your self-esteem. When there is rapport, you can share both good and bad times without tension.

The foundation of any relationship is trust. You believe that the other person will not betray, reject, or hurt you. A trustworthy person will handle your feelings with care in order not to hurt you. They will not reveal confidential information. You trust that a friend will be concerned about you, and stand up for you when you are not there. You trust that a member of your group will share the workload as well as the credit for a successful result. In return, you will value the other person's feelings as much as you value your own. You will act with concern for the other person's well-being. Friends should "be there for each other."

How Relationships Work: The Social Exchange Theory

The social exchange theory was developed by social scientists to explain the give and take in relationships. It takes an economic approach. Relationships have "costs" and "benefits." In order for a relationship to be effective, the benefits of being in the relationship must equal or outweigh the costs of the relationship. If you are paying a high "cost" for someone else's "benefit," you might feel that you are being exploited.

The Benefits and Costs of Relationships

Benefits of Relationships

- pleasure in spending time together
- social and emotional support
- intellectual growth through discussion
- excitement of new activities

Costs of Relationships

- time spent together
- emotional and social support provided
- disappointments and hurt feelings
- mental strain of disagreements

Common interests and shared experiences mean that the relationship will provide opportunities for you to spend time together and to have fun. People who care about you encourage you to try new things and to develop your skills. Talking about your experiences, your thoughts, and your feelings with your buddies enables you to expand your understanding of yourself. Debates about hot topics with trustworthy friends and classmates will help to develop your thinking skills and your ability to present your arguments. Listening to your family and friends discuss their lives and their beliefs helps you to develop empathy for others and expands your horizons.

Relationships are based on reciprocity, a two-way interaction requiring a *commitment* from both people. Both people must commit time and energy to the relationship. It is also important to put effort into understanding the other person. Not all of your relationships will be friendships with equal partners. As a student interacting with a teacher, an employee relating to your boss, or a team player relating to your coach, you must still play your part in the interaction. There will be times when you have to *compromise* in order to achieve your goals in the relationship. As friendships develop over time, you become more willing to give without always expecting something back.

Reflections and Connections

1. Explain how someone you know makes you feel comfortable being with him or her. How does this behaviour create rapport?

2. Has someone ever betrayed your trust? How did they behave? How did you react? What has happened to the relationship since then?

3. How are your relationships with your teachers different from your relationships with your friends?

CHECK Points

1. ✓ What is the meaning of rapport in a relationship?

2. ✓ Is it possible to be friends with someone if you do not share the same interests?

3. ✓ Describe the behaviour that indicates that a person is trustworthy.

4. ✓ How does compromising contribute to reciprocity in a relationship?

Types of Relationships

Family Relationships

Your first relationships are those within your family. Parents were probably the most important people in your life for the first few years and often continue to be most involved in your life. By custom and by law, parents share in most of the important decisions about your direction in life. Brothers and sisters may have been your earliest playmates, roommates, and fellow pranksters. Siblings who are close in age may become good friends, and older siblings might mentor and support younger brothers or sisters. Grandparents may provide advice as well as being your number one fan club. Your family relationships will change as you and the other members of your family mature and change.

Because you did not choose your family, these relationships are different from friendships. Although families may live together, they often have difficulty finding time to spend together. Family members may have known each other all their lives, but may not take the time to share each other's thoughts and feelings about their daily lives. Family members might also have difficulty recognizing that your needs and interests are changing as you mature. Some family relationships might not meet your needs. Because family relationships are not equal relationships, getting along with each other is more demanding. In the long run, however, family relationships can be worth developing since they will last a lifetime.

Family members share the important events of their lives with each other.

Friendships

Friendships develop outside of your family. They may take on different forms as your opinions about people change and as you get to know various people. Most of your friendships will be with your peers. **Peers** are those people who are of the same age group as you and with whom you share similarities. Friendships vary in the degree of commitment expected and in the amount of time and energy spent together. The closer the relationship you have with someone, the more time you want to spend together and the more you may open up to the other person.

Siblings may continue their childhood rivalries for a lifetime.

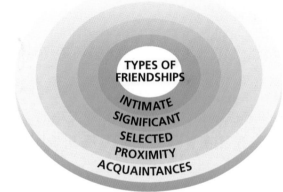

TYPES OF FRIENDSHIPS
INTIMATE
SIGNIFICANT
SELECTED
PROXIMITY
ACQUAINTANCES

Children begin by loving their parents; after a time they judge them; rarely, if ever, do they forgive them.

Oscar Wilde

CINDY, WOULD YOU PUT SOMETHING IN MY YEARBOOK?

SURE, ELWOOD, LEMME HAVE IT.

STMMP!

NEXT

It was fun being your friend. Cindy

Acquaintances

Most of the people that you meet in your daily travels are acquaintances. Acquaintances are people who you know because they participate in the same activities as you do. The students in your class, your school principal, your neighbours, and the cafeteria staff are some of your acquaintances. You probably know their names, and you might exchange casual greetings or chat about the impending snowstorm, but the conversation is usually brief. Acquaintances help fulfil your social needs because their familiar faces provide a sense of belonging in your school and community.

Proximity Friends

When you form friendships with students in your class, the people you work with, and the other woodwinds players in your school orchestra, you are forming proximity friends. Circumstances and common activities have thrown you together and given you common interests. You will share your ideas about situations and discuss other people in casual conversation. Perhaps you spend time discussing the difficulties of yesterday's homework and exchange stories about the other students in your English class. Proximity friends make shared activities more enjoyable but, when the term ends or you change jobs and you no longer have an activity in common, the relationship will probably not last.

Proximity friends make school activities more enjoyable.

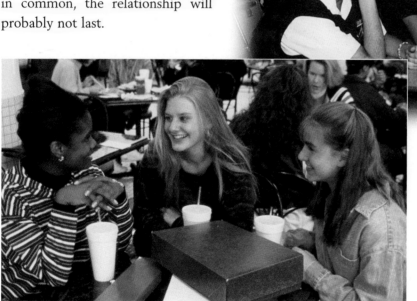

Selected friends choose to share their lives.

Selected Friends

When you arrange to get together with someone because you enjoy their company, you are becoming selected friends. These friendships are based on shared interests and some affection or concern for each other. You will share your opinions with your selected friends and exchange stories about your lives as you get to know each other better. You might have a circle of selected friends with whom you spend your social time at lunch, after school, and on weekends. These friendships often last a long time, sometimes a lifetime.

We helped each other and lived in the shelter of each other. Friendship was the fastest root of our hearts.

Peig Sayers

For many adults, their most intimate friend will be their spouse.

Significant Friends

Spending many hours on the phone talking together, sharing your inner thoughts and feelings in extended conversations is a sign that a friendship has become a significant relationship in your life. This sharing creates empathy, a better understanding of how the other person feels. In addition to sharing many good times, significant friends offer emotional support and can help you cope with difficult times. Significant friendships require that you be there to provide the support and encouragement the other person needs to cope with their lives. This give and take can be challenging at times. Many people describe the emotional rewards of significant relationships as love.

Intimate Friends

Intimate friendships are your closest relationships with those you love. Many people choose one or two people with whom to share their innermost feelings about themselves and their lives. Intimate friends enjoy spending time together and make sacrifices for the benefit of the relationship. Intimate friendships require the greatest commitment of time and energy. For this reason, the end of intimate relationships can be very painful for both people. Sometimes, intimate friends find each other sexually attractive, as well. This will add another dimension to the relationship as well as offering another way of being close. Although it is not always the case, in adulthood, most Canadians expect that their most intimate friend will be their spouse or life partner.

Breaking up is hard to do; teens need tender loving care to mend their first heartache—tips for parents of heartbroken teens

By Cindy Barrett

Although 14-year-old Krista McVeigh* had gone out casually with other boys, Matt was her first serious boyfriend. When he abruptly ended their relationship after six months, Krista was devastated. Her mother, Diane, a nurse, was upset to see her daughter in such pain. "I listened to her tears and her anger," she says, "and I tried to get her to voice how she had envisaged their future. In her heart, Krista realized that she and Matt had not been headed toward a lifetime together." But Krista was still convinced she had done something to cause the breakup.

To teenagers, there is nothing quite so racking, so all-consuming, as the first time their hearts are broken. A bittersweet milestone, it often occurs at the same time that adolescents would rather lock themselves in their rooms than discuss their emotions with their parents. Although they may naturally turn to their peers for comfort, an adult—whether it's Mom, Dad, a favorite aunt, or other trusted grown-up—can help them sort out their tangled feelings.

Generally speaking, girls and boys react differently to breakups, notes Marion Balla, director of the Adlerian Centre for Counseling and Education in Ottawa. Boys tend to deny their painful feelings and often get involved with someone new very quickly to regain their emotional balance. Girls often get stuck going over what they should have done and are insecure about starting a new relationship.

How do you reach a lovelorn teen?

- Do not minimize the pain. Stale advice such as "There are plenty of fish in the sea" shows your teen you are not taking her situation seriously, says Sue McGarvie, an Ottawa clinical sex therapist and specialist in adolescent development. A breakup, she says, "absolutely rocks their world." Let your teen go through all the regular stages of grieving, such as denial, anger, and acceptance, and be prepared to put up with more moodiness than usual.

- Offer an empathetic ear. "Restaurants are a good place to talk because you can really focus on each other," says Balla. "Allow your teen to volunteer information without badgering, interruption, or advice."

- Be available. "If you're ever going to take the day off work," says McGarvie, "this is the time." Take your teen to a movie, the museum, or shopping at the mall. Show him/her that he/she is still lovable with plenty of hugs. You may even want to tuck him/her in at night or make comfort food he/she liked when he/she was younger.

Even with the most compassionate care, some teens take the breakup badly and can wallow in the pain, although, depending on the relationship, most get over it within a month. If your child refuses to take part in his or her usual activities or is not functioning at home or school then seek help from your family doctor or guidance counselor.

"If you give your teen the time now, your bonds will strengthen in a significant way," says McGarvie. "They will be able to come to you when things are bad—and when things are good." In Krista's case, her mother was delighted that her daughter, always a private child, was prepared to tell her what had happened. "It was a turning point in our relationship," says Diane.

* The McVeighs names have been changed. Courtesy of *Chatelaine* magazine © Rogers Publishing Ltd. Reprinted with permission of the author.

Functional Relationships

We must learn to live together as brothers or perish together as fools.

Martin Luther King Jr.

Relationships that meet your academic, financial, or health needs, rather than your emotional needs, are called ***functional relationships***. You will have a patient-doctor relationship with your family physician and a customer-retailer relationship with the person at the milk store. At school, you have student-teacher relationships with your teachers and group partnerships to complete projects. These are all functional relationships that meet needs other than your emotional and social needs. Functional relationships are usually clearly defined. You are expected to behave appropriately to complete tasks or perform roles, regardless of your personal feelings about the other person.

Functional relationships are quite different from friendships. They can contribute to your social development by providing opportunities to work and co-operate with others. It is possible to develop a rapport with others in a functional relationship. A comfortable relationship with your Math teacher will help you learn how to solve equations, and treating all of the customers to a friendly greeting might earn you extra tips. Your success on group projects depends on forming effective working relationships. Developing positive relationships with all of the people in your life can be rewarding.

Winning the game depends on teamwork not friendships.

CHECK Points

1 ✓ Why are acquaintances important in people's lives?

2 ✓ Why do proximity friendships usually fade after time?

3 ✓ Why do people have fewer significant friends than proximity friends?

4 ✓ Why do people have so few intimate friends?

5 ✓ Identify examples of functional relationships in the lives of adolescents.

Rapport with your teacher can make the lessons easier to understand.

Reflections and Connections

1. Identify people in different areas of your life who are acquaintances. What needs do these relationships meet for you?
2. Who were your proximity friends when you were in Grade 7? What did you enjoy doing together? Where are they now?
3. Who have been your most significant friends so far? How did they affect your personal development?
4. Describe a rewarding functional relationship you have had. Would you call the person a friend?

Building Effective Relationships

As you have discovered, a developmental task of adolescence is forming mature, effective relationships. As you grow and mature, your peer relationships will become very important. Although relationships are necessary to help you on your journey through life, forming new relationships can be a challenge. Your relationships will become more varied as you get to know new people in a variety of situations. Whether starting a new school, being placed in a group for a project, joining a team, getting a job, moving or emigrating, you will have many opportunities to develop new relationships. These relationships can meet your developmental needs and provide lasting rewards. Some relationships last a lifetime, so it is worth investing time and energy to get to know people.

"To make a friend, be friendly" is straightforward advice! Look for opportunities to participate in activities that you are interested in and enjoy with other people. Introduce yourself to people that you meet. To develop

The formula for achieving a successful relationship is simple: you should treat all disasters as if they were trivialities but never treat a triviality as if it were a disaster.

Quentin Crisp

rapport, make other people feel comfortable by smiling and speaking in an open and friendly manner. Keep up to date on current issues that people are interested in so that you have something to talk about when meeting people. The more interests you have, the more you will have to talk about as you get to know new people.

Reciprocal relationships meet the social and emotional needs of both parties. Relationships develop over time, but first impressions provide the chance to attract someone's attention in a positive way. Avoid judging people too quickly; make your own decisions after you have an opportunity to learn about their

Keep up to date with current events and issues, so that you have something to talk about with friends and acquaintances.

interests and attitudes. To become friends, share your ideas and opinions and ask questions to find out about the ideas that others have. Be willing to spend time with other people and to compromise on activities. These considerate behaviours on your part will help to build a lasting friendship.

Building trust in a relationship requires that you take risks. Open and honest relationships are based on equal sharing. Open up to others and reveal information about yourself gradually. People become uncomfortable if you tell too much about yourself when you first meet. Observe how others respect your confidences. Of course, you will be expected to respect their confidences as well. As relationships become closer, people reveal more of their inner thoughts and feelings. You can control how close the relationship becomes by how much you reveal of your inner self and how much you ask the other person to reveal to you.

Relationships: A Valuable Part of Life

Relationships can provide rewarding companions on your journey through life. Your family gets you started on the journey, but you will continue to make friends and form new relationships throughout your life. Some of your friends will be temporary companions such as your best friends from Grade 2, the kids in your cabin at camp, the person on the same shift at your fast food job. Significant relationships may help you decide the direction in which you will live your life. Friends enable you to share your dreams, to share satisfying experiences, and to laugh and have fun.

CHECK Points

✓ How can you start talking to someone you are sitting beside at school?
✓ Why is it a good idea to reveal personal thoughts and feelings slowly in a relationship?
✓ How can you meet another person's social and emotional needs?

Reflections and Connections

1. Trace how one of your friendships developed over time.
2. Have you ever formed an inaccurate opinion of someone based on first impressions? What made you change your opinion?
3. Explain how reciprocity applies to several of your relationships.

Roles

Roles are the parts that you play in life. You play many parts at the same time—child, student, cousin, classmate, first soprano, cashier. Some roles are assigned to you by your position in the family, the community, and society. Others roles might be chosen. Each role has a set of behaviours and expectations associated with it. Students are expected to carry books to class, and to answer the teacher's questions. They may also be expected to whisper to each other, occasionally forget to do their homework, and to complain when a test is announced. Fitting in depends on learning how you are expected to behave and what behaviour you can count on from others.

Teachers and students have clearly defined roles in school.

career **Link**

What is a counsellor?

Counsellors work with people to help them with their personal problems and issues.

What does a counsellor do?

Relationship counsellors can be guidance counsellors, social workers or marriage counsellors. In each case, the counsellor interviews the clients to get their case history to determine the problems within the relationship. While guidance counsellors provide assistance with school-based relationships, such as teacher-student relationships, social workers often focus on problems within family relationships. Counsellors will provide clients with information, skills, and strategies that will help them achieve more effective relationships. They often consult with other professionals, such as psychiatrists, teachers, and physicians in order to provide the most effective approaches for their clients.

Where do counsellors work?

Most counsellors work within community agencies such as the school or in the social service sector. Counsellors also work in medical settings such as clinics and hospitals. Specialized counsellors, such as marriage counsellors might also maintain a private practice where members of the community can seek help on an hourly basis.

To work effectively to achieve a group goal on a school assignment, planning a social activity, or in the workplace, you need to take on a group role. Usually, each member of a group takes on a specific responsibility or task role. This role may be assigned to you, or the group might decide together what the task roles will be. Some people tend to take on the same task role every time, but you will become more skilled if you try different roles within the group. Sharing the workload requires that all participants work hard to complete the task by doing their share and that all participants work hard to complete the task by doing their share and by helping others

By working together, you can accomplish the task faster and better!

Are you an effective group member?

When working in a group do you...

- know your role in the group?
- do your fair share?
- give the project all of your creative effort?
- share all your ideas with the group—no holds barred?
- encourage the contribution of others?
- actively listen and participate in all group discussions?
- stay focussed on the group goal?

if necessary. In effective groups, members share their ideas freely and encourage quieter people to share by asking for their ideas and opinions. Finally, someone has to be the task master who ensures that you meet the deadlines, but everyone should stay focussed and avoid distracting the group from its goal. These behaviours define the role of effective group members.

Role Models

Role models are people who you respect and admire. They provide positive examples of how to play a part in your life. As you observe people in their roles, you learn the behaviours

Teenagers who take on leadership responsibilities can be important role models for younger children.

that are expected, as well as the appropriate attitudes. If you observe mom and dad making supper and doing yard work, you learn that both sexes can perform these tasks. Parents are your first role models. They teach you how to be male or female, and the attitudes and behaviours necessary to be an adult. Older siblings, senior students, even people in the media can be role models. Your challenge is to choose role models who will show you how you can live your life in the directions that you would like to go. There will be many roles that you will not choose. As you learn new roles, you will find it easier to behave as others expect you to and to fit comfortably into new situations in your life.

The expectations of the different roles that you play in life might conflict. Perhaps your parents expect that you will go directly home after school. Your friends expect you to go to the food court at the mall, and your teacher expects you to work with your partner to complete an assignment on the computer. At the same time a neighbour has asked you to baby-sit her son for two hours while she goes to an appointment. You may feel pressure to conform from several directions. Situations such as this one challenge you to clarify your priorities and make some decisions. Since you cannot be in all of the roles at the same time, you will need to negotiate with others. In order to make sure that you are meeting your developmental needs and maintaining your important relationships, you need to be clear about the roles that you play.

CHECK ✓Points

✓ List examples of roles that are assigned to teenagers.
✓ Identify examples of chosen roles.
✓ What behaviours are expected of a student at your school?
✓ How can roles conflict?

Reflections and Connections

1. List all the roles that you are playing at this time in your life. Classify them as either assigned or chosen.
2. In a chart, compare the behaviour that is expected in your role as son or daughter with that of a participant in a group. Share your chart with a classmate.
3. Create a list of the role models in your school and your community.

Summary

1. Relationships are necessary for your intellectual, emotional, and social development.
2. The challenge of family relationships is to allow them to change as people mature without losing the special closeness of being a family.
3. The different types of friendships vary in the amount of affection, trust, commitment, and sharing that each person brings to the relationship.
4. Functional relationships are based on performing your role, not on personal feelings.
5. Relationships develop gradually as you communicate with each other.
6. Roles describe the expected behaviour that is specific to a relationship.

In this chapter, you have learned to

- explain the importance of relationships for meeting social and emotional needs;
- describe the characteristics of healthy and effective relationships;
- explain how a variety of relationships meet different needs; and
- explain why people need to belong to and participate in groups.

Activities to Demonstrate Your Learning

1. Using social science research methods, conduct surveys to determine the characteristics of an

effective friendship for adolescents. Analyze the results. Summarize the qualities of a good friend and explain how these qualities meet social and emotional needs. Using your conclusions, design a checklist for individuals to use for becoming a better friend.

2. Design a poster advertising the qualities of effective group members to post in your classroom.

3. Write a personal reflection describing a relationship with a significant adult who has been a role model for you. Explain how the relationship enriched your life.

4. Using social science research methods, design a research study to determine which roles are most important for adolescents at your school and to identify the expectations of each role. Report the results of your research on a chart.

5. In small groups, conduct an investigation of the portrayal of relationships in your style of popular music. Classify the qualities that are associated with different types of relationships. Compile and report the results in a song in the style your group has chosen.

6. In pairs, draw caricatures depicting the roles of various family members, such as mom or older brother, from the point of view of a child. Post the caricatures for everyone to see. Discuss how your ideas about those roles have changed now that you are a teenager.

Enrichment

1. Analyse the various relationships of a character in a novel, and explain how each met or did not meet the character's developmental needs.

2. Compare the portrayal of friendships in current and old television programs to determine whether ways of relating have changed. Present your results as a chart.

3. Write a short story that describes the progression of a relationship from first meeting to significant friendship.

Communication Is the Key!

Think about the important relationships you have in your life. What do you need from your various relationships? How do you communicate your needs to those who are important to you? Is communication easy or strained? Do you have arguments with significant others? Is it easier to talk to some friends than it is to others? Does your communication differ according to the role you are in?

By the end of this chapter, you will be able to

* define the process of communication;

* identify and use effective communication skills to send and receive messages;

* demonstrate the use of good verbal and non-verbal communication strategies;

* describe assertiveness within the context of good communication; and

* identify the importance of written communication within a technological environment.

You will also use these important terms:

active listening	feedback
assertive	I-messages
bias	non-verbal communication
body language	passive listening
communication	prejudice
communication filters	verbal communication

What Is Communication?

Communication is the exchange of information between two or more people. It involves a process of creating and sending messages as well as receiving and interpreting messages. Poor communication can cause much confusion for all concerned. Effective communication can enhance both situations and relationships.

For communication to take place, there must be a sender and a receiver of a message. In the role of the sender, you are required to compose a message and then transmit it to another person. In the role of the receiver, you must first hear or see the message and then interpret or decode the message. Good communication happens when both the sender and the receiver arrive at a shared meaning about the message. Both the sender and the receiver must have certain skills to communicate effectively.

Writing a letter or e-mail message is a form of verbal communication.

The Communication Process

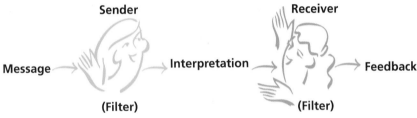

Sender   Receiver

Message →   → Interpretation →   → Feedback

(Filter)   (Filter)

Sending Messages

Speech is the mirror of the soul; as a man speaks, so he is.

Publilius Syrius

Within the communication process, the sender must be skilled in creating a clear and accurate message. Sending a message also involves transmitting those words in a manner that ensures clear and accurate interpretation. This involves knowing appropriate vocabulary and terminology for verbal communication. **Verbal communication** involves the use of words. When you speak to a person or a group, you are using verbal communication. When you write a letter or an e-mail message you are also using verbal communication.

Sending a message to ensure appropriate interpretation can involve the transmission of words using non-verbal communication. **Non-verbal communication** uses elements other than words to convey a message. These elements may include gestures, eye contact, body movements, or facial expressions. Non-verbal communication can be used on its own or in conjunction with verbal communication. Tone of voice or word emphasis can completely change the way in which the message is communicated.

What are you saying?

For each of the following phrases, emphasize the bolded word. Note how the message takes on a slightly different focus in each case.

WHAT do you want me to do?

What **DO** you want me to do?

What do **YOU** want me to do?

What do you **WANT** me to do?

What do you want **ME** to do?

What do you want me to **DO**?

You can reveal a great deal about your feelings through your use of body language. *Body language* is a form of non-verbal communication. For example, if you are sitting slumped in a chair, staring out the window when someone is talking to you, you may be demonstrating feelings of boredom or disinterest. A great deal of body language happens without thought or planning. Sometimes the non-verbal messages of body language do not match the words in your verbal communication. When this is the case, you are sending "mixed messages."

When the eyes say one thing, and the tongue another, a practiced man relies on the language of the first.

Ralph Waldo Emerson

Sometimes a simple gesture is all that is required to send a message. A touch on the shoulder can communicate love and affection.

Here are some forms of body language. Identify the messages that they are communicating.

You may be using words to engage in verbal communication, but your body language is saying that you are not really involved in the communication.

Tips for Sending Effective Messages

Think before you speak.	Take into account the points that you want to make before speaking. This may save you from embarrassment or hurting others.
Express a positive attitude.	Try to send a message that is warm and enthusiastic whenever possible. No one really likes to hear the complaints or criticisms of others.
Send clear and specific messages.	Ensure that your thoughts are organized in meaningful way. Try to use detailed facts to support the point that you are making. Use clear language.
Speak clearly.	Use words that others will understand. Make sure you pronounce words correctly and distinctly. Do not talk too slowly or too quickly.
Be aware.	Be aware of all aspects of your communication so that both verbal and non-verbal communication are sending the same message.
Check for listener understanding.	Be aware of the listener's body language for information about how your message is being received. Ask for verbal feedback if you have any questions about the non-verbal feedback you are receiving.

Receiving Messages

Receiving messages within the communication process also requires specific skills. Listening to a message is as important as sending it. The way you listen affects the quality of communication. When someone hears your message without really taking in the meaning, they are using **passive listening**. How do you feel when someone you are speaking to does not respond? Silence is often frustrating. You do not know if your message was received accurately. People who use passive listening frequently miss the purpose or the value of a message.

> *For one word a man is often deemed to be wise, and for one word he is often deemed to be foolish. We should be careful indeed what we say.*
>
> Confucius

Active Listening Can Enhance Other Aspects of Your Life

Gather more information about your world.	You can acquire a greater knowledge about your environment, current events, and the people around you. Having more information increases the number of tools you have to ensure clear communication.
Promote good relationships.	When you demonstrate to others that you really listen to their messages, they feel valued and understood. These feelings of acknowledgement provide a solid foundation for caring relationships.
Focus on others and be less self-involved.	Focusing on others allows you to practice empathy. You will grow as a person when you are able to appreciate differences in others.
Show interest in others.	When you show interest in the speaker's message, you show interest in the speaker. When people are made to feel worthwhile, it boosts their self-esteem.

When both speaker and listener observe good communication methods, interactions are pleasant and effective. Being aware of all aspects of communication will make conversations flow more easily.

CHECK ✓ Points

✓ What is the difference between verbal and non-verbal communication?

✓ Why is it important to be aware of your own body language?

✓ What skills are necessary to send effective messages?

Reflections and Connections

1. Have you recently had an uncomfortable communication with someone? Describe their body language during that conversation. What message was the non-verbal part of their communication sending?

2. Describe some gestures that you use on a regular basis to send messages to others.

3. How does your tone of voice change when you are sending positive messages? negative messages?

Active listening involves concentrating on what is being said so that you understand and remember the message. Part of the communication process is providing *feedback* to the person you are speaking with. Nodding, smiling, or making eye contact indicates that you are listening and understanding. Verbal responses such as "I understand" will encourage communication to continue. A question or short statement might clarify your understanding of the message. As an active listener, you will be able to demonstrate that you both heard the message and understood its meaning.

How to Be a Better Listener

- **Concentrate.**
  Focus only on what the other person is saying. Eliminate distractions whenever possible; for example, remove your earphones. Stay in the present and do not think ahead to what your response will be.

- **Listen with a purpose.**
  Recognize the reason for the communication. Your friend might be telling you about her problem because she needs to express her feelings.

- **Keep an open mind.**
  Be prepared to accept and honour the other person's point of view.

- **Connect to the speaker.**
  Make eye contact. Lean toward the speaker. Keep your facial expression open and interested to encourage the flow of information.

- **Interpret the message.**
  Make sure you are interpreting the message correctly by using the feedback mechanisms of active listening. Keep your own ideas and your feelings about the speaker from getting in the way.

- **Do not interrupt.**
  Sometimes it is very tempting to speak out in the middle of someone else's message. Interruptions, however, can confuse the speaker's message and make the speaker feel less valued.

- **Be positive.**
  Assume a positive attitude in your role as the listener. A positive attitude will help to keep you and the speaker motivated throughout the communication process.

- **Control your emotions.**
  If what you hear affects you emotionally, stay calm and continue listening to the speaker. Express your views and feelings after you have heard the message with an open mind.

✓ Describe the difference between active and passive listening.

✓ What are the benefits of active listening? When is active listening required?

✓ Explain three ways to improve your listening skills.

Reflections and Connections

1. Describe a recent situation when the person you were talking to used passive listening. What were the consequences? How did you feel?

2. How do you feel when someone interrupts you? How do you respond when you are interrupted?

3. What are some strategies that you could use to control your emotions during a conversation?

Communicating Assertively

To be able to ask for what you want and express your feelings directly you need to be **assertive**. When you are assertive, you communicate your ideas and feelings firmly and positively. The ability to be assertive is a strong confidence booster. It demonstrates self-respect and your ability to be fair to yourself and others.

Assertiveness is not aggressiveness. Aggressive behaviour is negative and can be forceful and destructive. Passive behaviour does not engage people emotionally. Although passive behaviour literally involves doing nothing, it can be negative as well. Assertive behaviour is positive and does not put others down. Communicating assertively allows you to communicate negative emotions without hurting others. Learning to be assertive will enable you to express your needs and wants without guilt or apology. Being assertive enables you to stand up for what you believe in without attacking or hurting others.

CASE STUDY

Joanne was worried and angry when Stephanie did not show up at the library after school. Stephanie had promised to be there at 4:30 P.M. She had not called to say she could not make it. "You are so inconsiderate," Joanne told Stephanie the next day. "You should have called me. You made me waste my time by waiting around for you. Why are you not more dependable?"

*Joanne could have expressed her feelings more effectively by using an I-message. Her messages were a direct attack on Stephanie. You-messages are fre-*quently used in communications. They are more negative and can create adversarial confrontations. You-messages tend to accuse and blame others for your negative feelings. I-messages describe and explain the speaker's feelings without put-downs. Using I-messages helps to keep negative emotions under control and not complicate a message.

Using an I-message, Joanne could have said, "I was really worried when you did not show up at the library yesterday. I was afraid that something had happened to you. Besides that, I also wasted an hour of my time waiting." Joanne's I-message would allow Stephanie to focus on the situation without feeling the need to defend herself.

Once you have made this type of clear non-judgmental statement, you can go on to state what it is that you want. Stephanie can apologize and offer to call next time, or Joanne could add, "Would you please call me next time if you are not going to make it?"

Summarize in own words

Using I-Messages

A tool for clear and assertive communication is the use of I-messages. ***I-messages*** allow you to take responsibility for how you feel. They are non-threatening and help the other person to not feel blamed or belittled. You will find communication with others much easier when you start to use I-messages.

I-messages have three parts:

- **"I feel** (name the emotion you feel)
- **when you** (state the specific behaviour)
- **because** (explain what effect the behaviour has on you)."

Once you have made this type of clear non-judgmental statement, you can go on to state what it is you want.

Roadblocks and Bridges in Communication

Roadblocks in the communication process can result from the use of you-messages. The person receiving the message might feel attacked and feel the necessity to defend themselves. When this happens, communication breaks down since the receiver cannot focus on the communication and defend themselves at the same time. I-messages can be used as bridges in communication in order to prevent breakdowns in the process.

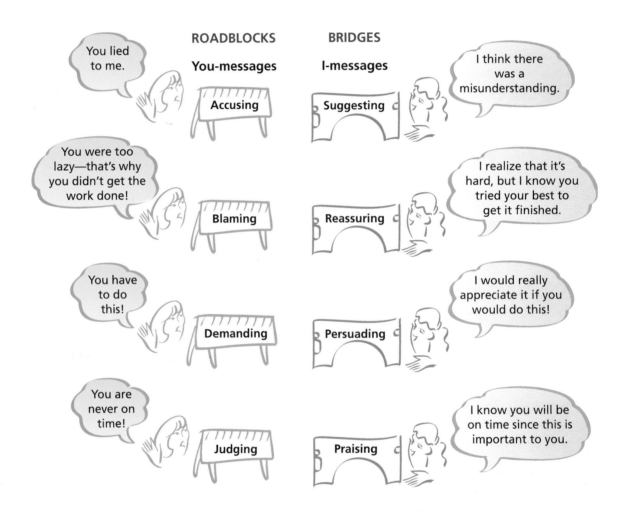

ROADBLOCKS | BRIDGES

You-messages | **I-messages**

You lied to me. | Accusing | Suggesting | I think there was a misunderstanding.

You were too lazy—that's why you didn't get the work done! | Blaming | Reassuring | I realize that it's hard, but I know you tried your best to get it finished.

You have to do this! | Demanding | Persuading | I would really appreciate it if you would do this!

You are never on time! | Judging | Praising | I know you will be on time since this is important to you.

Detours in the Communication Process

Sometimes messages are unclear due to a variety of **communication filters**. It can be difficult to determine if the content of a message is fact or opinion if any one of these filters comes into play.

Filters include such factors as bias, stereotyping, and ethnocentrism. **Bias** is a way of looking at things from a very narrow perspective. Often biases are based on opinion and not fact. Stating that "Men are better athletes than women" shows a bias against women that is not based on fact.

Prejudice is a type of bias against an individual or group who share certain characteristics. When the characteristics of some group members is generalized to all group members, then stereotyping takes place. Making statements like "All people of Oriental heritage are good at Math" is an example of stereotyping. Ethnocentrism is a type of bias where individuals view people from cultures other than their own as having less importance or value. When any of these types of filters come into play during the communication process, messages can be unclear and misinterpreted.

There are many ways that the communication process can be enhanced. Building bridges to maintain the clarity and flow of messages is important.

Positive Strategies

Timing   Effective communication takes place only when the sender and the receiver are ready to focus on each other's messages. Selecting the right time to convey a message may make a difference in how it is received. Knowing when someone is ready, willing, and able to listen takes skill and sensitivity. Using empathy will help you to determine how well the receiver is prepared to listen to your message.

Honesty   When complete openness and honesty exist between two people, communication is greatly enhanced. Both parties can be sure that they are receiving all of the necessary information needed. They can also be certain that they have information about how the other person is feeling.

Humour   Communication can be fun. Using your sense of humour in communications can make a message more lighthearted and pleasant. Humour can help to put a more positive spin on a message. Always make sure that you do not make jokes at other people's expense or to the detriment of the message. Laughing at yourself can show maturity. While frequent "joking around" may be inappropriate, in some circumstances humour can be a good way of coping with stress.

CHECK Points

✓ What is the difference between bias and stereotyping?
✓ What roadblocks to communication can be avoided by using I-statements?
✓ Describe why timing is an important factor in communication.

Reflections and Connections

1. Describe a recent communication where you could have been more assertive. Create an I-message to practice what you might have said.
2. How can you use humour in communication?
3. Describe a communication that you have had or witnessed recently where bias or stereotyping was involved. Create some I-messages that might be useful to use in this type of situation.

Cultural Distinctions in Communication

Communication patterns, symbols, and meanings are different the world over. Cultural differences can lead to significant misunderstandings between individuals and groups.

Cultural Differences

New Zealand	Maori tribe members touch foreheads as a greeting instead of shaking hands.
China	A letter written in red ink gives the message that the relationship between the writer and the reader is being broken.
Some Asian countries	The common North American hand gesture for calling someone is used only for dogs.
Some African countries	It is not polite for a man to speak to his mother-in-law, so he must ask someone to relay the message for him.
Vietnam	It is the custom to indicate that you have heard a question, then proceed to answer the question. This causes confusion when translated into English as, "Yes. No, thank you."

Written Messages

Although the spoken word is the most common form of personal communication, the written word is a valuable communications tool in many ways. Newspapers, magazines, bumper stickers, billboards, posters, and letters are all popular and effective ways of transmitting messages. Poetry, short stories, and novels provide a way of communicating information about the arts, history, and culture.

From a personal perspective, you use the written word to send messages by letter or e-mail to your friends. You communicate your knowledge and understanding of what you learn in school through your written assignments. You send notes and cards to friends for birthdays and other special occasions. You may leave messages for family members to give them important information. Your message may be a memento for someone and become a part of their permanent records.

From a career perspective, many jobs require effective writing skills. You may be required to write reports, produce orders, or compose business letters. You may also be expected to communicate with colleagues via e-mail messages to keep up-to-date.

All of these personal and professional written communications require specific skills to ensure that messages are conveyed effectively. Like clear speaking, concise writing ensures that your messages are understood. When writing, select clear precise words, avoid biases, and use I-messages whenever possible. Writing messages certainly takes longer than speaking, but as a method of communication, it allows for more time to compose, consider, and reflect on the message you are sending.

Above: Written messages can be special mementos that are cherished for a lifetime.

Below: Effective written messages in the workplace are very important for corporate communication.

Communications Technologies

Our society is in a very strong information age. The communication of information is important in all aspects of peoples' lives. We need up-to-the-minute information about many areas of our lives. It is critical that we can send and receive messages quickly, efficiently, and clearly. Technology is developing rapidly to increase the pace and the quality of communication. While technology provides us with a large amount of information in a small amount of time, it is also creating new opportunities as well as challenges for individuals and families.

career Link

What is a writer?

A writer is a person who writes for a profession or business as an author, copywriter, or journalist.

What does a writer do?

Writers can use their skills and training in many areas. Creative writers such as novelists, playwrights, and poets conceive and write material for publication or other public presentation. Technical writers analyze material such as specifications, notes and drawings, and write manuals, user guides, and other documents. Copywriters study and determine selling features of products and services and write text for advertisements and commercials.

Where do writers work?

Writers may specialize in a particular subject or type of writing. Technical writers might work in the pharmaceutical or medical field, while copywriters might become speechwriters for a political party. Most writers work in the service industry; however, many work in the business and education sectors. Publishers account for most of the employment in the freelance sector. Freelance writers work within the freedom of their own homes; however, they often do not have the security of a steady income.

Telephone and Wireless Communication

The telephone provides us with quick and easy verbal communication. It gives more information than the written word as you are aware of the sender's tone of voice and the pace and pitch of their speech. Many non-verbal cues are missing with telephone communication. Facial expressions, gestures, and posture are all non-verbal cues that are unavailable through telephone communication. Yet, many people spend more time talking to their friends by telephone than face-to-face. With the increased use of cellular phones, you can be in touch with friends and family no matter where you are. Cellular phones allow parents of teens to always know where their children are.

Above Left: The use of cellular phones can lead to unsafe situations if they are not used properly.

Above Right: You and your friends often talk on the telephone even after just seeing each other in person. Phone conversations are often extensions of important personal communications.

Technology for the Disabled

For many people communicating is difficult. Technology has made great strides in assisting people with disabilities to communicate. Many new devices have provided mechanisms for greater personal, social, and financial independence.

Technological Devices

Voice synthesizer This device is used with special software to convert words on a computer screen into verbal messages for the visually impaired. They can respond by using a Braille keyboard to type their responses.

Talking glove This communication device is used by people who cannot speak. The person makes letters and words by sign language while wearing the glove. Sensors within the glove translate the sign language into speech. A microphone with a computerized voice, relays the message to the listener.

Computers

Computers and the Internet have significantly changed the ability to send and receive messages and information. A person can access information from around the world in seconds. This information can be sent in the form of a message to another person in the same amount of time. The nature and the accuracy of information available on the Internet is sometimes questionable. It is important to use sound judgment and decision-making skills to evaluate the quality of the information and messages received via the Internet.

The new electronic interdependence recreates the world in the image of a global village.

Marshall McLuhan

Computers are rapidly changing the way we send and receive messages. How have your communication patterns changed as a result of computer use?

CHECK ✓Points

✓ Why is it important to be clear when writing messages to your friends?

✓ Describe situations in the workplace where the clarity of written messages would be very important.

✓ List some reasons why telephone communication may be preferable to written communication.

Reflections and Connections

1. What factors do you consider when writing a card or other important message to a friend?

2. How do you use the computer for sending and receiving messages? Describe how it fits into your personal communication pattern.

3. Describe why you do or do not like communicating by telephone.

Summary

1. The communication process requires skill in the sending and receiving of messages.

2. Effective messages are sent using well thought-out, clearly worded I-messages through verbal communication strategies.

3. Messages are effectively received through active listening that uses a feedback process.

4. Non-verbal communication involves body language and tone of voice. When a person's non-verbal communication contradicts their verbal communication, the result is a mixed message.

5. Knowing the potential roadblocks in the communication process enables the use of skills such as assertiveness and humour to build bridges to good communication.

6. Technology has a great impact on communication today. Computers and other information technologies are providing new methods of communication for people with a variety of abilities and needs.

In this chapter, you have learned to

- define the process of communication;
- identify and use effective communication skills to send and receive messages;
- demonstrate the use of good verbal and non-verbal communication strategies;
- describe assertiveness within the context of good communication; and
- identify the importance of written communication within a technological environment.

Activities to Demonstrate Your Learning

1. Draw a lifeline to show your life. Draw a vertical line to show your ages from birth to the present, beginning with birth at the top of the page. Describe ways that you used to communicate at different stages in your life. Identify the problems you have experienced in communicating at these stages. How has your growth and development helped in solving communication problems?

2. Survey your classmates or another group of friends to identify what types of communication they have the most difficulty with. Using the information you have acquired, create a pamphlet to provide communication tips for adolescents.

3. During one class period, observe and record the various forms of non-verbal communications demonstrated by your classmates. Beside each form, identify the message communicated.

4. Pretend you are defending your point of view on an important issue. Describe examples of how you would use I-messages to effectively communicate your point of view.

5. Role-play a variety of situations where poor communication has caused a problem. As a group, identify the barriers to good communication and discuss ways in which the communication could be improved.

Enrichment

1. Research how communication has changed over the past 100 years. Describe the impact of these changes on individuals and families.

2. Watch a television show or movie and pay particular attention to a character who is trying to communicate with someone else. Identify and describe how the character is using body language and other non-verbal communication. Analyse how the non-verbal communication is helping or hindering the clarity of the message.

3. Keep a communication journal in which you record examples of effective and ineffective communication that you observe. Evaluate the examples of communication using the communication process.

4. Search the Internet to determine the types of jobs that require good communication skills. Investigate the need for both verbal and non-verbal skills. Compile your results in the form of a report.

Managing Your Relationships

Do you enjoy your interactions with other people in your life? Relationships can be a source of enjoyment and happiness. Relationships can also result in conflict, disagreements, fighting, and unhappiness. Relationships that are not meeting your needs can make you feel that you are being taken advantage of. Some people are even hurt physically by someone they know. What can you do to maintain healthy relationships with others?

By the end of this chapter, you will be able to

✺ describe strategies for maintaining effective relationships in your life;

✺ demonstrate problem-solving skills for improving relationships;

✺ demonstrate conflict-resolution and negotiation skills; and

✺ suggest strategies for dealing with annoying, harassing, and abusive relationships.

You will also use these important terms:

abusive relationship	mediation
acquiescence	negotiation
bullying	peer pressure
compromise	problem solving
discrimination	self-awareness
harassment	tolerance

Healthy Relationships

Relationships are very important to most teenagers. Unfortunately, relationships are also the number one problem for adolescents. Healthy relationships require effective communication skills and a clear sense of what you want from each of your relationships. Taking charge is the key to having relationships that enrich your life.

Healthy relationships are based on giving and taking to meet the needs of both people. If you are being expected to give more than you want to, or if you are making excessive demands on someone else, the relationship will probably result in conflict. Problems can arise in a friendship, a family relationship, or a functional relationship at work or in school. If you are not comfortable in a relationship, you can decide whether to try to improve the relationship or to end it.

Decisions about your relationships should be based on clear expectations about the relationship and how you feel about the other person. You need to understand what you want from a friend, a sister, a group member, a teacher. You must also know what you are willing to give to the relationship. You can decide what type of relationship you want or need with each person in your life. Being sensitive to how you think and feel about yourself is the first step to taking charge of your relationships.

If you treat people right, they will treat you right— ninety percent of the time.

Franklin D. Roosevelt

Self-Control

In order to take charge of the relationships in your life, you must first develop self-awareness and self-control. **Self-awareness** means that you are attuned to what you value or believe as well as what you feel. Self-control means that you can express those beliefs and feelings appropriately. You are learning to identify your feelings better as you mature. Remember that your feelings belong to you and that you can control how you express them. You can choose how you will respond when you have strong feelings within a relationship. Developing skills for expressing your feelings appropriately can help you to have healthy relationships.

Emotional Hijacking

Sometimes emotions can swamp your brain and prevent you from thinking clearly. There is good reason for this. The human brain developed from the bottom up—the feeling brain developed before the thinking brain. When you respond to a situation, your emotional brain triggers a release of hormones that makes you act—before you have time to think! This is great in an emergency, but can cause you to overreact in other situations. The old advice about counting to ten before you act can help you relax and give you time to think.

Respect

Good relationships begin with mutual respect. Respect means seeing someone as a worthwhile person. Liking someone depends on shared interests or values. You do not have to like people to respect them. People with different points of view or interests usually consider their opinions to be just as right as you consider yours to be. *Tolerance* demonstrates that although you have clearly made up your own mind about your beliefs and interests, you respect the rights of others to have different beliefs and interests. Treating others with respect can often prevent conflict.

Tolerance

Tolerance is the ability to accept differences. Tolerance requires that you accept that you have chosen your opinions, beliefs, and interests as your own point of view. You must also recognize that others may have different opinions, beliefs, and interests. Tolerance does not require that you agree with others, only that you agree to disagree. Tolerance is a sign of intellectual, emotional, and social maturity.

Overreacting to situations can threaten your relationships.

A disagreement with a friend can make you question the relationship.

Balance

When you feel uncomfortable about a relationship, set aside your feelings and reflect on the relationship. Ask yourself whether there is a balance in the relationship. Does the relationship meet the needs of both parties? Do you feel pressured to act in a way that makes you uncomfortable? Decide whether you are willing to change your behaviour in order to maintain the relationship. Sometimes, you might decide that you cannot devote time, energy, and emotion to the relationship anymore. On the other hand, you may decide that the relationship is valuable and work with the other person to solve the problem in order to maintain the relationship.

Peer Pressure

Peer pressure is an attempt to influence your behaviour by those who are in a similar age or interest group. Peer pressure can be positive or negative. If someone is encouraging you to do something helpful or academically challenging, that pressure is positive. Pressure from your friends that discourages you from doing something wrong is also positive. Negative peer pressure often comes in conflict with your own wants, needs, or values. Some teens drink alcohol, smoke cigarettes, use illegal drugs, or get involved in sexual relationships because of negative peer pressure. A good way to determine if peer pressure is positive or negative is to determine whether the activity is dangerous or destructive in any way. If peer pressure is negative, you need to use strategies such as assertiveness and conflict resolution to pursue your own best interests.

Anger Management

Anger is an emotion that often occurs in conflict situations. If you feel that something is not fair, out of your control, or that you have been wronged in some way, anger is a natural response. Anger is not always harmful. It can motivate people to change unacceptable situations. Handling anger in a useful way can produce positive results. Experiencing anger is normal and is likely to occur in relationships where you spend a lot of time with a person.

Anger is a brief madness.

Horace

Problems arise when anger is expressed in a harmful way. If anger is expressed by physically or verbally attacking others, conflict situations become worse. Keeping anger inside can also have a negative effect. Unexpressed anger to avoid dealing with a negative situation can lead to stress, depression, or physical illness. You can learn to manage your angry emotions in a way that allows you to express them yet handle them in a positive way. It is important to express anger in ways that are not hurtful to others.

Steps in Managing Your Anger

- Learn to forgive and forget. Realize that everyone makes mistakes and needs to have a second chance.

- Focus on any positive or humourous aspects of the situation.

- Talk about the situation to a close friend of family member. Talking it out can often reduce the intensity of your angry feelings.

- Identify the ways in which you can manage the situation. An effective strategy might be to diffuse your anger, punch a pillow, or channel your energy into an activity such as painting or running.

- Pinpoint the cause of your anger. It might help to write down what you are feeling and why.

- Acknowledge that you are angry.

Speak when you are angry and you will make the best speech you will ever regret.

Ambrose Bierce

CHECK ✓Points

✓ How do the expectations of friends, siblings, and group members differ?

✓ Why is it good advice to "count to ten" before you act?

✓ What is the difference between positive and negative peer pressure?

Reflections and Connections

1. What are some strong feelings that can affect your relationships with other people?

2. Why might it be difficult to respect someone who is different?

Conflict in Relationships

Some conflict is a normal part of relationships. Conflict is a struggle between people who have opposing points of view. It can involve individuals or groups such as families, friends, community groups, or even whole countries. In principle, conflicts are not bad. Conflict can be useful when it forces people to address issues that require attention. Conflict can provide the opportunity to see a new perspective or come up with an innovative new idea. Conflict is usually not pleasant, but it does not have to be painful. Learning to handle conflict can actually help to strengthen relationships.

When two people have differences of opinion, their friends often get involved in the conflict. Choosing sides may be difficult when you understand and empathize with both individuals.

Types of Conflict

Conflicts can be internal. Suppose your family wants you to attend your grandmother's seventy-fifth birthday party. You have always been close to your grandmother and would like to share this special event with her. The conflict arises when your best friend's family decides to have a surprise sixteenth birthday party for her on the same night. Your family would like you to be at your grandmother's party, but they tell you that it is your decision. You want to please your family, but at the same time you want to be there for your best friend. In this case, the struggle or conflict is within you. The battle is between your own strong feelings.

External conflicts involve others when the wants, needs, or values of one person or group clashes with those of another. There are several causes for external conflicts between individuals and groups.

Power Struggles

Power is the ability to influence and control people or situations. Power struggles occur when there are issues that are important to both sides. Many arguments concerning power issues occur in families with adolescents. Both parents and teens want control. An example is when parents want you to clean your room and you feel that since it is your personal space the standards of cleanliness should be within your control.

Personality Differences

Each person has a unique style, along with a set of values and traits that form their personality. These differences can enhance and enrich relationships, however, the differences can also create conflict. Suppose two boys shared a bedroom. One liked to read and study in a quiet environment, while the other worked best with loud rock music. What potential conflict lies in their differing styles? Often skills in *compromise* will prevent long-term conflicts in this type of situation. If the boy who likes to listen to rock music while studying agreed to use headphones, the other boy would be able to study in the quiet environment that he needs. Dialogue, empathy, and compromise are the tools necessary to solve problems with personality conflicts.

Situational Conflicts

Many situations can cause sudden and heated arguments. Anywhere people interact—at home, at work, or in recreational environments—conflicts can ignite. In these cases, the stronger the emotional connections between people, the stronger the conflict. Amanda is going to meet her boyfriend, Jake, for lunch. When he is not in the spot that they had agreed upon, she goes down the hall to look for him. She finds him talking to Lisa. Amanda gets very upset and runs away.

Situational conflicts can be intense, but are often short-lived. A discussion, where listening skills and empathy are used, can often dissipate a situational conflict. In Amanda's case, she listened to Jake's explanation. He had been talking to his lab partner, planning for their biology presentation. They needed to make sure that they were prepared and their planning had taken longer than anticipated. He had not realized what time it was and felt badly about missing his meeting time with Amanda. Amanda had thought that Jake was starting another relationship. She had jumped to conclusions. Often, situations are not what they seem. A cool head and time for discussion will frequently avoid serious conflicts.

CHECK Points

- ✓ Where can conflict create a positive situation?
- ✓ What are the differences between internal and external conflict?
- ✓ How are power struggles different from situational conflicts?

Reflections and Connections

1. Consider a conflict that you have experienced recently. Did it have a positive or negative result?

2. Think about an internal conflict that you have had. How did you deal with it? How was it different from external conflicts you have experienced?

3. How do you feel when you do not have power in a situation? What type of situation might you be in where a lack of power would cause a conflict?

How to Resolve Conflicts

When conflict occurs, there are many ways that you can respond. It is important to consider many factors before you take action. Consider your relationship with the other person. Always be concerned for your personal safety and well-being. Realize that walking away from an explosive situation is usually a positive choice and not a sign of cowardice. If the other person is someone you will probably never see again, walking away may be the wisest act.

In situations where conflicts arise within a relationship, you will want to use a positive process to resolve the conflict. When conflicts within relationships remain unresolved, the relationship often deteriorates. With a satisfactory resolution, the relationship can continue to be strong and grow. If the other person is someone that you want to maintain a personal relationship with, you may want to dissipate the conflict through communication and conflict resolution.

How to Identify the Early Signs of Conflict

Look for communication clues.
- name-calling
- yelling
- threats
- insults
- argumentative language

Leave the scene if there are clues that the situation may become physical.
- pushing or shoving
- slapping or kicking
- threats with weapons

No matter how a conflict arises, you can participate in a process that will resolve conflicts in a peaceful way. *Negotiation*, the process of settling a conflict through *problem solving* and co-operation, is a tried and true approach.

Help for Selecting a Solution

Acquiescence: When you change your behaviour because someone asks you to, that is ***acquiescence***. In some situations, you may discover that your behaviour is causing a problem for someone else that results in conflict. If the specific behaviour is not necessary for you, or if your behaviour could easily be changed, you might agree to change in order to solve the problem and resolve the conflict. Acquiescence can strengthen a relationship if it means meeting the other person's needs without sacrificing your own values or needs.

Compromise: If the people in conflict are fairly close to agreement, they may agree to compromise. This involves each person giving up something that they wanted in order to settle the conflict. This is a give-and-take method, where each disputant gives a little but maintains much of their positions. This solution helps disputants to express themselves within the context of the disagreement. Each person must recognize the value of the other person's opinions and feelings. Compromise protects the relationship and helps to maintain each person's self-esteem.

Negotiation: When disputants communicate to send and receive messages in order to reach a solution, they are in the negotiation process. This process is a purposeful face-to-face discussion with the intention of reaching a solution. We often hear about the negotiation process in labour conflicts. In the negotiation process, both parties must send clear and accurate messages. Both parties need to understand and carefully consider messages before responding. The negotiated solution should be satisfactory to both parties. With negotiation, both parties are committed to reaching a workable compromise. They share the goal of working together in order to achieve agreement.

Mediation: Sometimes people looking for a solution to a conflict are not successful. A deadlock may occur when disputants cannot agree on any resolution. In ***mediation***, an unbiased third party is asked to help reach a solution that both sides can live with. Mediators are neutral. In their role, they ask questions and are careful listeners. Keeping both parties calm and productive is a mediator's goal. Mediators help disputants see solutions that those in conflict are too upset to see.

Using acquiescence, compromise, negotiation, and mediation will assist in selecting the solution to the problem. Make sure that everyone agrees with the final solution, otherwise the conflict cannot be resolved. Working together to resolve the conflict can strengthen a relationship and enable you to face future issues with confidence.

The Negotiation Process

1. **Define the problem.**

   Make sure that everyone involved understands exactly what the problem is. This can be done by having each person describe the problem from his or her point of view. It is important to show respect to each disputant at this time. Sometimes conflicts are easily resolved at this stage as it may clarify that, in fact, there was no conflict at all. Perhaps there was just a misunderstanding or a communication problem.

2. **Identify ownership of the problem.**

   Everyone needs to understand who the problem affects. The person bothered by the situation owns the problem. If the problem disturbs more than one person, ownership must be shared.

3. **Identify possible solutions.**

   Each person tries to come up with as many creative solutions to the problem as they can. Everyone should feel free to speak openly without being criticized or judged. Good listening skills are important at this stage of the process.

4. **Evaluate each suggested solution.**

   Each disputant identifies the parts of a solution that he or she agrees with or cannot accept. Solutions can then be modified to be more acceptable, or perhaps to be discarded as alternatives.

5. **Select the best solution.**

   This is the hardest part of the process. Often there is no best solution. Many options are available that come with an acceptable solution in this part of the negotiation process.

6. **Check to see that the solution is working.**

   If any of the disputants are not satisfied with the arrangements, you need to go back to the previous step. Go through the steps again, until a mutually acceptable solution to the conflict can be found. It might be necessary to get help to find a solution.

7. **Deal with failure.**

   If a satisfactory solution to the conflict is not possible, you may have to agree to disagree. The best solution may be the acceptance of disagreement and respect for differing opinions. This might also change or end the relationship. It is important for all parties to acknowledge the disagreement and ensure that the unresolved issue does not create future conflict.

Reflections and Connections

1. Describe a situation where it would not be worthwhile implementing the strategies of conflict resolution.
2. Negotiation is a technique that is often used by adolescents and their parents. List some ways that negotiation can be helpful in conflicts between parents and teens.
3. What are some strategies that can be used when conflict resolution has failed?

Unhealthy Relationships

Relationships that cause distress or hurt are unhealthy relationships. Behaviours that are unhealthy in relationships are harassment and abuse. Although unhealthy relationships may be painful, it is often difficult to avoid them. Preventing these behaviours at home, at school, and at work is best for everyone. Unfortunately, you may have to deal with unhealthy relationships that you witness or that you experience.

Harassment

Harassment is unwelcome and uninvited attention that causes the victim distress. Harassment creates an environment that is stressful for the victim. Harassment sometimes results in discrimination when people are excluded from participation. Harassment can create the uncomfortable feeling that being different may make you a victim. To stop harassment, it is important to recognize the forms it can take.

CASE STUDY

Simone's Story

High school was an ordeal for Simone. She was the only student at her school who dressed in a particular non-conformist style like others who listened to the music she liked. She was pushed and shoved in the halls. Someone took her Science textbook and wrote obscenities in it before it was "found" in the classroom. When she sat down at a table in the cafeteria, the others got up and left. During gym, other students aimed the ball at her head. Her locker was sealed with crazy glue. Simone was hospitalized for depression.

Gopaul's Story

Gopaul's family moved in December to his suburban high school. He wants to make new friends. Some of the students are making fun of him, asking why he has a handkerchief on his head, and why he has a "girly bun." Gopaul tried explaining that Sikh men do not cut their hair and that he will wear a turban when he is a man. Still the boys laugh at him. One boy even tried to grab his hair to remove the comb that holds his hair in place. Gopaul wishes he could go back to his old school.

Franco's Story

Ever since Grade 5, Franco has been harassed by the same group of kids. They called him names such as "fairy." He was beaten up by several boys after gym class. He was suspended for fighting with the other boys in his class. In Grade 9, they wrote on his locker and chanted quietly every time he walked by in class or in the halls. Franco's father told him to toughen up and be a man, and his mother asked that he be transferred.

Bullying

Bullying is cruel behaviour directed at specific individuals. Usually the bully and the victim know each other but are not friends. There may be a power difference between them. Bullies may act alone or in groups. Sometimes one person acts while others just watch. There are several types of bullying. Physical bullying is much easier to recognize, but all forms of bullying are hurtful. The emotional scars of bullying can last a lifetime—both for the victim and for the offender.

Types of Bullying

Physical Bullying Hitting, kicking, taking or damaging property, or blocking someone's way are some of the ways that physical bullies act. This type of bullying is most visible.

Verbal Bullying Name calling, insulting, constant teasing, and threats are examples of verbal bullying. Some forms of verbal harassment are so common that we dismiss them as "just a joke" and often do not pay attention to the hurt caused by the harassment.

Relational Bullying Convincing peers to exclude someone, or spreading rumours about someone are forms of bullying that are least visible.

Racial/Ethnocultural Harassment

Racial and ethnocultural harassment involves physical, verbal, or relational bullying directed at others because of their race, skin colour, religious beliefs, ethnic origin, citizenship, or country of origin. It is meant to exclude them because they are members of a group that is different from that of the bully. Telling racially-based or ethnic jokes, making fun of where someone was born, or making stereotypical insults are forms of harassment. Refusing to work with someone or excluding someone from your group are forms of *discrimination*. Racial and ethnocultural harassment hurts and angers others and damages relationships.

Physical bullying is more visible than verbal and relational bullying.

Sexual Harassment

Sexual harassment includes unwelcome sexual comments, sexual advances, and requests for sexual favours. Unwelcome comments include judging or making fun of someone's body. Displaying sexually offensive pictures or graffiti, telling offensive jokes, and inappropriate staring are also sexual harassment. Sexual discrimination, which means treating people unequally because of their gender, sexuality, or sexual orientation, often goes along with sexual harassment. Both boys and girls can be the victims of sexual harassment.

Students Discuss Attack on Sikhs

By Jim Wilkes, Staff Reporter

Students at an Oakville high school, troubled by accusations of racism in an attack on classmates last month, are confronting issues of racial intolerance.

More than 1300 students at Oakville Trafalgar Secondary School met in assemblies yesterday to discuss multiculturalism, racism, and ways it can be fought on campus and in the community at large.

"In our school and in all sectors of society, it's important to recognize that we live in a multicultural society," principal Larry Di Ianni said in an interview. "We need to tolerate the fact that we have different groups and we need to go beyond that to a higher degree of acceptance and oneness."

"That was the theme of the assemblies: How do we move to oneness where as individuals we respect differences?' he said. "I think the kids reacted very positively to that."

The assemblies were originally scheduled to mark International Day for the Elimination of Racial Discrimination but also became a forum to discuss a Feb. 15 snowball attack on two Sikh students that ended in six arrests.

Halton police say the incident, which began with two students being pelted with snowballs as they left school, escalated into racial slurs, punches, and kicks.

Police said one Sikh student suffered a fractured skull and another had his turban torn off and thrown

into a garbage can. Three other students who came to their aid, two of whom are Sikh, were also assaulted, police said.

Eight students received school suspensions ranging from five to 20 days.

"Any time you have a significant incident that galvanizes opinion, it provides opportunity to move on and reach a greater degree of understanding," Di Ianni said.

"I see a very healthy beginning to a new and deeper understanding."

Five teenage males and a teenage female are charged with assault in the incident. They are to appear in Oakville youth court April 5.

The Toronto Star, March 22, 2000
Reprinted with permission – The Toronto Star Syndicate

Is It Really Harassment?

Harassment is often excused by saying, "I was only teasing." Teasing is only possible in a respectful and affectionate relationship. Your parents might tease you as a way of offering criticism of your behaviour in a loving way. Friends also tease sometimes. Teasing can easily become hurtful, so teasing must be done with care. Remember, harassment is unwelcome and uninvited attention. If you ask someone to stop the teasing, and the behaviour continues, it becomes harassment.

You should be able to tell the difference between friendly behaviour, flirting, and harassment. The key words to remember are "unwelcome" and "uninvited". Healthy relationships, even those that are just developing, are based on reciprocity. People give and take on an equal basis as they get to know each other. You have the right to try to get to know someone better. You also have the right to indicate that you do not want the relationship to go further. You have the right to ask someone to stop what they are saying or doing if it makes you feel uncomfortable.

Harassment is often excused by saying, "I was only teasing."

Reflections and Connections

1. Racial and ethnic jokes are a form of harassment. Explain.
2. Is it harassment to tell a racial or ethnic joke if there is no one of the race or ethnic group mentioned present? Why or why not?

Take Action to End Harassment

Harassment is not only a problem for the victim. Harassment is also a problem for those who witness it and for the community. Harassment creates an environment of fear. The victim may be afraid to go to places where they might be harassed. Those who witness harassment may be afraid that they will be the next victim. The offenders continue to have problems dealing with their aggression and may become more aggressive.

People need to feel safe from harassment to be able to learn, work, and enjoy their lives. To prevent harassment of any kind, there must be no tolerance of any type of harassment at school, at work, or amongst your peers. Some schools have a "zero tolerance" policy with clear consequences for any form of harassment. You will develop your own principles for caring about and respecting others.

If you are being harassed, or if you witness harassment, act to stop the behaviour. Speak to the people who are bothering you individually. Speak calmly and firmly. Tell them that the behaviour is harassing you and you want them to stop. You can speak to the person by yourself, or you might feel more comfortable having a friend come with you. Make a note of your complaint and record the time and details in case you have to report the harassment. This type of assertive communication might be enough to end the harassment.

If the harassment persists, or if you do not want to confront the offender yourself, tell a trusted adult what is happening. Speak to a teacher at school,

CHECK Points

1. ✓ Define the types of harassment.
2. ✓ Distinguish between harassment and discrimination.
3. ✓ What is the difference between flirting and sexual harassment?

In the end we will remember not the words of our enemies, but the silence of our friends.

Martin Luther King Jr.

CASE STUDY

Action Plan: I-Message

Petra had been the victim of sexual taunts from several boys in her class for several days since the beginning of school. She felt very uncomfortable sitting near them in class or walking by them in the hall. Once, when she told them all to stop, they laughed and said they were just teasing. She decided to confront them with an I-message and a clear request that they stop the behaviour.

Taking a friend to stand by her for courage, she spoke to each boy separately when she saw him without his friends. This is what she said:

"When you tease me by commenting on my body or asking me to do sexual things, I feel very uncomfortable and I cannot concentrate on my work. You are harassing me, and I want you to stop your harassment now." Two of the boys looked uncomfortable, but agreed to stop. The third boy told her to "lighten up." This is what she said to him: "Your comments are not funny. I do not want to hear them. They are harassment. If you continue, I will report you to the teacher for sexual harassment." The boys stopped their comments.

You might feel safer making an anonymous call about harassment.

a counselor, a family member, or call an agency. Reporting harassment is not "snitching" or "tattling." Reporting is done to solve a problem and prevent harm. Harassment hurts people and should be stopped so that other people are not hurt. Ignoring behaviour suggests that it is acceptable and encourages offenders to continue.

Bullies and other offenders should stop their harassing behaviour. Social science research has found that bullies usually turn to new victims. Bullies are more likely than other people to commit criminal acts as adults. Reporting harassment is the only way offenders will get help to change their aggressive and hurtful behaviour. Do not tolerate harassing behaviour when you witness it, so that everyone will feel more secure. Support the victims, not the offenders, by speaking up.

CHECK Points

✓ What is the purpose of I-messages in stopping harassment?

✓ Why is it important to record the details of harassment?

✓ Distinguish between snitching or tattling, and reporting.

Reflections and Connections

1. Describe a situation in which you witnessed harassment. How did the victim behave? How did you feel?

2. Have you acted to stop harassment? What did you do?

3. Does your school have a policy regarding harassment? What is it? What are the consequences for harassing behaviour?

Abusive Relationships

Some relationships can cause you physical or emotional harm. If you are being hurt in a relationship with a family member, a boyfriend or girlfriend, or someone that you trust, you are in an *abusive relationship*. Physical abuse includes hitting, slapping, and other physical harm. Emotional abuse can include demeaning comments, threats, and insults that diminish your self-esteem. Although it is human to lose control occasionally or to say something that hurts someone, it is abusive to do these things as a regular pattern in a relationship.

Sexual behaviour may also be abusive. Sexual touching and sexual contact are sexual assault if you do not want it to happen. It may also be sexual assault even when you have agreed to the contact if the person has power over you because of their age or authority. Sexual assault is a serious offense. In legal terms, no sexual contact is okay unless you give your permission as an equal partner in a relationship.

If you are being abused in a relationship, you should speak to someone you trust about what is happening. This is an important step towards ending the abuse. You may feel that you love the person and feel afraid of losing the person's love. It can be scary to point a finger at an adult or someone who is admired by your peers. This fear is not a reason to endure physical or emotional abuse, or unwanted sexual contact. If you are under 16, your situation will probably be reported to Children's Aid authorities who will investigate the circumstances and determine how best to protect you. Your relationship with the person who is abusing you changed forever when the abuse began.

CALL IF YOU'RE BEING OVERWORKED AT SCHOOL.

CALL IF YOU'RE BEING WORKED OVER AFTER SCHOOL.

KIDS HELP PHONE
1 800 668 6868

THERE'S A LOT ON THE LINE

Kids' Help Phone is a national organization that will listen to you and suggest how you can end the abuse.

Ending a Relationship

Sometimes relationships do not work out. Sometimes people who have been friends grow in different directions by developing new interests, or by changing beliefs about important issues. People find themselves drifting apart over time, much as they drifted together in the past. This is a normal part of social development. It is also a normal part of your social life to find that someone you were close to has grown apart and that the relationship has faded. By choosing to spend less time together, by sharing fewer of our experiences with someone, and by revealing less of our inner self, we can choose to distance ourselves from former close friends.

Romantic relationships also end. Because of the high degree of sharing in these intimate relationships, we place higher demands on each other. When we expect so much, it is more likely to feel that the relationship does not meet our needs. It is difficult to explain why two people are attracted to each other. It is just as difficult to understand why the attraction may fade. It is fair to the other person to let them know that your feelings have changed.

The Green-Eyed Monster that Eats Love

By Kelley Teahen

"Something is wrong. Something is wrong with her daughter Kate."

That's the opening line of a short story, Back Pain, by the late Bronwen Wallace. It's about a mother, Barbara, who is worried about Kate since she's started dating Danny, a teenager who "really is charming. So maybe she's making too much out of the way her daughter's face seems to be operated by remote control, fading out every once in a while, then coming up again when her eyes flick to Danny's."

At the end of the story, Barbara finds her daughter curled up and weeping in bed, with a black eye and a bloodied nose. "Did Danny do this to you? The mother asks, knowing there is no other explanation. She crawls into bed with her daughter and holds her until her weeping and shuddering stops.

Teenage romance isn't supposed to be about bruises and blood. But, in our popular culture, jealously is supposed to be a sign of someone's love. Green-eyed jealousy and its black-eyed cousin, control, are becoming partners in far too many teenage relationships. Because once you can't control through emotion or intimidation or fear, then you start trying to control with fists.

Jan Richardson, executive director of Women's Community House, is a hard woman to shock, after all her years of organizing emergency response for the victims of our domestic wars. But earlier this year, even she was shocked. She looked around the shelter and realized she saw teenagers everywhere. "We had more teens in the residence that week than we'd ever had before."

UWO psychology professor David Wolfe specializes in studying dating violence in teens. He's found many teens in violent dating relationships come from families where there's been violence between mom and dad at home.

Just like kids from loving families learn to look for healthy love, many kids from violent families stick close to the script they know. It doesn't help, Wolfe says, that the script featuring jealousy and control shows up in many of the fictional love stories of movies, TV, and music videos.

In his work, he provides alternatives for such kids in their early teen years.

"We steer them clear by discussing what a healthy relationship is. They hear from people who have been in abusive relationships and hear that their partner had been jealous.

"They need good examples. How do you solve problems without becoming hostile or controlling? They need an awareness that jealousy is not the best way to express your concern or love for someone."

There are other factors: unrealistic expectations about the degree of closeness that should exist between a couple; a teen's off-kilter view of what a relationship is supposed to provide; and what's considered "normal" within a teen's group of friends.

Dangerous Patterns

While there's a volatility and drama in most teenagers' relationships, Wolfe points out there are certain patterns that teens and their parents, should recognize as dangerous. Does he not let her pursue her own interests and friends? Does she obsessively try to control his every move? Does he put her down and criticize her in front of other people?

The boyfriend in Wallace's story was all charm and smiles around adults, but was critical of his girl-friend, Kate, in front of other teens, a fact her mother found out only because Kate's brother started bugging her when she wouldn't eat her pizza at dinner one night. "'It's because of Danny isn't it,' Jason cuts in. "Last night, when he said you looked like Lois Campbell in that new skirt you had on.'"

For both boys and girls, teasing can be a playful way to communicate and negotiate the swirling waters of attraction. But it doesn't take long for teasing to become bullying. The barbs become a one-sided attack, vicious in spirit instead of playful.

Some kids will brush up against this kind of relationship and manage to get themselves out. I had one high-school friend whose boyfriend wanted to control her every move, including, and especially, what she wore. It escalated when she moved to our school from another city, because he couldn't watch her every moment. But she learned something about avoiding that kind of dynamic and went on to far healthier relationships as she grew up.

The ones who have more problems are the ones Wolfe describes as "needy kids." They haven't had much love or affection in their lives and, once they've found what looks like love to them, they'll hold on to whomever, whatever, they can. They'll put up with a lot. At first, a little jealousy is a small price to pay for having a boyfriend. It might seem even, well, romantic.

But respect, not jealousy, is at the heart of true love. And it's a shame some of our teens end up finding that truth only after they see themselves in a mirror with a battered face—and sometimes, not even then.

The London Free Press, 02-14-1998

An I-message allows you to express your changed feelings in a clear way without hurting the other person. Breaking up will cause pain, and you might experience grief for a wonderful experience which has ended. When a relationship ends, both people should eventually retain their self-esteem as worthwhile individuals who are worthy of being loved again.

Summary

1. Healthy relationships meet the needs of both parties without causing harm to either.
2. Healthy relationships are maintained through emotional self-control and respect for other people.
3. Peer pressure can encourage conformity to positive behaviour or to negative, destructive behaviour.
4. Conflicts can be resolved by communicating effectively to find a solution that meets the needs of all parties.
5. Harassment is unwelcome attention that causes distress. Assertive communication often stops harassment.
6. Young people who are being abused in a relationship should seek help from a trusted adult.
7. When a relationship is no longer working for you, end the relationship without demeaning the other person.

In this chapter, you have learned to

- describe strategies for maintaining effective relationships in your life;
- demonstrate problem-solving skills for improving relationships;
- demonstrate conflict-resolution and negotiation skills; and
- suggest strategies for dealing with annoying, harassing, and abusive relationships.

Activities to Demonstrate Your Learning

1. Develop a series of slogans which express strategies for ensuring that relationships are healthy based on the criteria for healthy relationships described in this chapter.
2. Gather examples of problem relationships from personal experiences, observations, and television programs. Write a case study in the form of a letter to an advice column. Exchange case studies with a partner and write a letter in response, recommending the problem solving that can improve the relationship.
3. Write a case study describing a conflict involving an adolescent. Using role-playing, demonstrate several effective conflict-resolution strategies. Decide which strategy is most effective.
4. In pairs, role-play assertive communication to stop various forms of harassment.
5. Using social science research methods, conduct a survey to determine the forms of harassment that have been experienced by students in your school. Analyze the results of your survey. Write a report recommending the action that should be taken by students to end the harassment.

Enrichment

1. Read a novel about an adolescent. Analyse whether the character managed his or her relationships effectively.
2. Compile a collection of terms, proverbs, and traditional advice concerning the management of relationships. Evaluate the wisdom of these expressions. Develop a glossary of the most valuable "words of wisdom" to distribute to your classmates.
3. Develop a brochure for adolescents in your community identifying the resources available to those needing help with abusive relationships.

What Is a Family?

Your family was your first environment. The members of your family will be life long references on your journey through life even if you do not see them for years. What is your family like? Is your family the same as those of your friends? How do you spend time with your family? How has your family affected your life and development? Your memories of family life from your childhood will be the standard against which you compare family life forever. What is a family and what should a family be like?

By the end of this chapter, you will be able to

❇ explain several definitions of family;

❇ summarize the universal functions of families and their effects;

❇ describe the variations in family forms and relationships; and

❇ describe various points of view on the responsibilities of families.

You will also use these important terms:

adoptive family	functions of the family
blended families	goods and services
breadwinner family	lifestyles
common-law	lone-parent families
culture	nuclear families
dual-income family	nurture
extended families	siblings
family	socialization
family form	spouse
foster family	stepparents

What Is a Family?

The family is one of nature's masterpieces.

George Santayana

If you were asked to draw a picture of a family, what would it look like? What would the family be doing? You are most likely to picture a family just like the one you grew up in. Or you might draw a stereotype based on the image of families on television. You might think of a family in terms of the individual people: Maria, Sean, and Ian. You could describe a family in terms of the roles people play, such as a mother, a father, a son, and a daughter. There are many different ways of defining family—perhaps all families are unique. Similarities and differences in families are evidence of the freedom you have to join together to create a family that meets your needs in your own way.

Who is a family? If there are so many varieties of families, why is it necessary to have a definition of family at all? A clear definition enables other people and organizations to determine whether people are related, and how they are related. Do people have to live together to be family? Will your family still be a family when the children grow up and move out of the family home, or if your parents get divorced? Do people have to be related by blood or by marriage? Are people who live common-law related to each other? Answering these and other questions helps to define "family" for two important purposes: defining family relationships makes it possible to identify who has the rights of being a family member, and who must bear the responsibilities.

Who a family is, how a family feels, and what a family does—this is the essence of a family.

Letty Cottin Pogrebin

How does it feel to be in a family? According to an Angus Reid poll in 1994, nine out of ten teenagers agree that their family lives are happy and full of love. We have seen that teenagers value their friends more than they value their families. This is normal because you are probably wanting to distance yourself from your family in order to leave them someday. In fact, most teenagers predict that family will be the most important thing in their lives when they are adults. This is not surprising since Canadians expect to marry their very best friend! Most teens believe they will fall in love and marry once for life and 84 percent expect to have children someday. How you form a family and live as a family when you are ready will vary as much as families do today—perhaps more.

Although it may not seem that way every day, most Canadians, of all ages, are generally satisfied with their family lives.

Definitions of the Family

What families do is the basis for the definition of the family used by the Vanier Institute for the Family. They define the *family* in Canada as "any group of two or more persons, who are bound together over time by ties of blood or mutual consent, birth and/or adoption and who, together, assume responsibility for the functions of families." This definition describes families as enduring, not temporary, so that they can do what families are expected to do. It recognizes that some people make a commitment to share their lives without legal papers and that some people are parents without having given birth. In short, according to this definition, if it acts like a family, it is a family.

On the other hand, Statistics Canada defines the family as "A now-married couple or a couple living *common-law*, with or without never-married children, or a lone parent of any marital status, with at least one never-married son or daughter living in the same residence." For the purposes of the census, it is essential to count people only once, so they must have a rule that people can only belong to one family on Census day. In fact, many people belong to several families at the same time. A man could have a family with his wife and child, and belong to his parents' family as their son. Individuals and organizations create definitions of the family to clarify the meaning of "family" for their own purposes.

CHECK Points

✓ How does the Vanier Institute for the Family define the family?

✓ What does the dictionary definition assume?

✓ Does the definition provided by the Vanier Institute for the Family fit your family?

✓ Why does the Statistics Canada definition specify "living in the same residence"?

Family what is it?

family n., pl. –ies. (1) a group of people related by blood, legal, or common-law marriage, or adoption. (2a) the members of a household, esp. parents and their children. (b) a person's children. (c) a person's spouse and children. (d) (attrib.) serving the needs of families (family doctor). (3a) all the descendants of a common ancestor. (b) a race or group of peoples from a common stock...

The Canadian Oxford Dictionary

Reflections and Connections

1. Identify some other definitions of family. Consider the definitions used for "family day pass," "family pack," or "family memberships."

2. What are some government, business, or community organizations that need a definition of "family" or "relative?"

3. In your opinion, which is the most important aspect of family—who it is, how it feels, or what it does?

4. Should there be a standard definition of family for all purposes in Canada? If so, what should it be?

The Universal Functions of Families

What do families do? Anthropologists have attempted to define the responsibilities of families by observing families in many different societies. They have concluded that families exist in all societies and are expected to take responsibility for functions that are not the responsibility of any other group. These *functions of the family* are universal. All societies have the same expectations, but how families perform these functions varies from place to place and from generation to generation. Each family performs the universal functions in its unique way.

Caring for Families

Families are expected to provide physical care for all family members. Families provide the necessities, such as food, clothing, shelter, protection, and medical care, to meet the physical needs of parents and children, the young and the elderly. Your family may also provide the non-essential wants which make your life more pleasurable, such as dessert, television, a bicycle, or a haircut by a professional stylist. If someone has no family, or if the family is unable to provide the physical care required, other ways of obtaining the support must be found or family members will suffer.

FUNCTIONS OF THE FAMILY

- physical maintenance and care of family members

- addition of new family members through procreation or adoption

- socialization of children for adult roles

- social control of children

- production, consumption, and distribution of goods and services

- nurturance and love

Adapted from Shirley Zimmerman, 1988

The Children's Aid Society

The Children's Aid Society of Toronto is one of the largest child welfare organizations in Canada. It is an incorporated not-for-profit agency governed by a volunteer board of directors and funded by the Province of Ontario.

Our legislated mandate is to protect children from harm. Because of our belief in the importance of early intervention, we also provide assessments, crisis intervention, counselling, and services to prevent child abuse and neglect. In addition, we help vulnerable communities protect and support their children. Many of these programs are offered in partnership with other community agencies.

Although it is our goal, whenever possible, to keep children with their own families, we offer a broad variety of substitute care programs, including foster care. These are provided for those children who, for various reasons, cannot remain at home or live with relatives. We also facilitate the adoption of Crown wards (permanent wards of the province).

The families and children with whom we work come from many racial, ethnic, and cultural backgrounds. As an organization, we are committed to non-discrimination and are striving to provide services that meet the unique needs and languages of minority groups.

Childrens' Aid Society, Toronto, March 2000. Reprinted with permission.

Family *lifestyles* vary depending on the family priorities and the income and resources available. Maintaining the home and caring for family members accounts for most of the household work of families. Families differ in how they share this workload amongst the family members. Some families choose to pay someone else to do some of the work for them. For example, parents might pay a day care provider to care for the children while they work, and pay someone at the restaurant to prepare dinner and clean up after the meal. Whoever does the work, ensuring that dependent children and the elderly are properly looked after is the responsibility of the family.

An autoworker sells his labour to the manufacturer in return for income to buy the things his family needs and wants.

In the past, families would provide for their needs directly, by producing their own *goods and services*. They grew their own vegetables, made their own bread, sewed their own clothing. Now families earn income to purchase goods and services. This results in a shift from providing directly to providing indirectly for the needs of the family. Families decide which members go to work to earn money. They decide when to spend the family money to buy the goods and services they need. They decide where to buy, choosing from businesses that sell goods and services. In return, many families earn money to fulfil their needs by providing the labour that businesses need. By working and spending, families play a vital role in the national economy.

Raising Children

Families in all societies fulfil the responsibility of adding to the population through having babies or adopting children. Traditionally, this also expands and strengthens the family. The question often asked of newlyweds, "When are you going to start a family?" emphasizes the common belief that children are necessary to form a family. By producing and raising children, your family has maintained a family line. They also enabled some people to achieve their goal of becoming grandparents. There is, so far, no substitute for men and women producing children.

New members of families become the next generation of society. For most people, the biological drive to reproduce in order to maintain the human species is usually taken for granted. However, the future of our society and our economy depends on population growth. Imagine a society in which few people chose

Families provide the next generation by having babies.

career **Link**

What is a family physician?

A physician is any qualified, legal doctor of medicine.

What do family physicians do?
Family physicians look after the holistic needs of the families within their practice. They usually give medical care to all family members including annual physicals and tending to the occasional physical disorders that family members may encounter. They treat diseases, physiological disorders and injuries of their patients. They evaluate the health of their patients through diagnostic procedures and often consult with other medical practitioners. They often provide advice to families about the best ways to manage their health care.

Where do family physicians work?
Most family physicians work in private practice. Many also work in hospitals and community clinics. Some physicians work in large medical centres as members of larger medical teams.

I do beseech you to spend more time preparing the youth for the path and less to preparing the path for the youth.

Ben Lindsey

Twenty-four percent of teenagers say that cultural heritage is very important to them.

Reginald W. Bibby and
Donald C. Posterski

to have children—which jobs would be lost? Children grow up to become consumers and workers who support the society's economy, and the citizens who support the social structure. All over the world, forming a family to raise children is often a major life goal.

Once families have children, they are expected to prepare them to leave the family and take on a meaningful role in adult society. They do this by a process called socialization. *Socialization* means teaching children the skills, values, and attitudes they need to be prepared for life. The skills, values, and attitudes will vary from family to family, but will reflect the general values and attitudes of the society in which the family lives. Families shape a child's personality and develop self-esteem by sharing good times and taking pride in children's accomplishments. Children learn to work, form relationships, and contribute to the family and to society.

Families pass on the *culture* of the family and the society. All families in all societies meet the basic needs of their members, but differ in how those needs are met. Your cultural background may stress buying or making certain foods that identify you as belonging to this group. Whether you eat pizza or samosas; whether you wear jeans, a kilt, or hijab to school; or whether you pray at a mosque, church, or temple in your community, your choices affect and are affected by what is available. Your family has taught you certain cultural values, such as respect for the elderly or not eating meat. Your culture may determine the expectations which you have learned concerning acceptable behaviour for your gender and for your age. Cultures rely on families to pass on their values, customs, and traditions to young people.

Love, Honour, and Cherish

Many families meet the emotional needs of their members by providing unconditional love because of who they are, not what they do. Family members are usually interested in each other's lives and maintain long-lasting relationships with each other. Families can *nurture* their children's emotional development by sharing their affection and companionship. Some parents, however, have problems of their own and have difficulty meeting the emotional needs of their children. As teenagers, many of you turn to your family for emotional support because you know that your parents will be there for you—even if they yell or get angry first. If your needs have not been met, realize that you can learn how to heal wounds and get a new start with your own future family. A loving family frees you to let down your guard, trusting that you are sheltered from the stresses of the world.

Family loyalty results from the unconditional love that many families provide. Socialization within the family teaches children respect and admiration for their parents, siblings, and extended family. Therefore, young children do not want to disappoint their parents. Teenagers are influenced by the behaviour of their family members. Whether they decide to follow or reject family values affects the reputations they will acquire. Often, adults are motivated by pride to behave well to maintain their family honour. Feeling happy and proud about your family encourages you to be your best to protect your family's reputation.

Families are expected to teach and enforce the social behaviour expected by the society. Most people behave themselves because they have learned to, not because they want to avoid arrest. Children learn acceptable behaviour

Young people learn their cultural traditions from their families.

CASE STUDY

Stephen's Story

As Stephen searched in the refrigerator for the chicken and vegetables mom had bought for dinner, he thought about what he had learned in Family Studies class today. His family was unusual in several ways, he thought.

Stephen had been adopted when he was a baby. When he was seven years old, his mom had a baby, his sister Becky. He lived with his mom and dad in an apartment downtown, but some day his parents wanted to move to a smaller town and buy a house.

Stephen's mom works at the library so she often works in the evenings and on weekends. His dad was out of work for quite a while before he started work at the car plant that opened back in the fall. Stephen often has to look after Becky. She looks up to her older brother, and he has fun hanging out with her.

Stephen enjoys high school. His parents have always expected him to do well in school, to set an example for Becky, so he hopes he will get good reports. Just last night, he had talked to his parents about what he wanted to do in the future. They reminded him that although they had to sign the form, he was responsible for choosing what courses he would take.

Lately, Stephen has been thinking about getting a part-time job, but he is not old enough. A job might interfere with playing on the soccer team, but he wants a new jacket. There is not enough money to replace his jacket this year because it still fits him, but it would be nice to get a better jacket. He does not get any allowance, so a job would also let him buy CDs once in a while.

As he turned on the stove to begin cooking dinner, Stephen decided that in their own way, his family was managing well.

What do you think?

1. What are the six functions of the family?
2. How does Stephen's family perform each of the functions?
3. Are there any functions that his family does not seem to perform?

and positive values from families that set limits in a loving environment, one that also criticizes and disciplines unacceptable behaviour. By law, parents are responsible for making important decisions for their children until they turn 18 years of age. In some places, parents must pay for damage done by delinquent children. These are ways in which families are expected to control the behaviour of family members to maintain order within a society.

CHECK Points

✓ How do families provide physical care for children?

✓ In what ways do families contribute to the economy of Canada?

✓ What actions do families take to socialize their children?

✓ How do families control the behaviour of individuals in their family?

Variations in Family Forms

Families come in many shapes and sizes. Families are expected to fulfil the same functions, but each family is formed individually by its members. By defining families by what they do rather than who they are, each unique family is free to focus on how it feels it can effectively meet the needs of its members. The *family form* is determined by the number of adults, the nature of the relationship between the adults, and the number of generations.

The most common image of a family is probably the *nuclear family* that consists of two parents and one or more children living together. The children are *siblings* to one another. The children can be the biological offspring of their parents, or they might have been adopted. In the 1950s, the most common nuclear family was the *breadwinner family* that had a single income earner, the father, and a stay-at-home mom. The *dual-income family*, where both parents are employed, is now the most common nuclear family in Canada.

Some nuclear families in Canada are *blended families*. Blended families, also called recombined or reconstituted families, have parents who divorced their first spouses and remarried. They have formed a family that includes children

Reflections and Connections

1. What are the necessities that families should provide? What are the luxuries, the "wants," in your opinion?

2. What goods and/or services do your parents create on the job in order to earn income for the family?

3. Which do you consider to be the most important function of the family? Would business choose the same function as most important? Why?

4. Which functions does government assume are most important? What evidence is there for your opinion?

5. What would happen to the economy if many Canadian couples decided not to have children?

from one or both previous marriages or from the remarriage. Children in blended families have *stepparents*. Maintaining contact with the previous spouses because of the children's relationships with their other parents can create a complex and challenging set of family relationships.

A lone-parent or single-parent family consists of one parent and one or more biological or adopted children. *Lone-parent families* are formed when a parent was never married, when a spouse or partner dies, or when parents separate and divorce. A child might be part of two lone-parent families if divorced parents share custody of the child. In most lone-parent families, the lone parent is a mother, but about one in ten lone parents are dads. If two lone-parent families join, they become a blended family.

Extended families consist of parents, children, aunts, uncles, grandparents, and other blood relations, living together or not. Most people have an extended family, but the amount of contact with extended family members varies. Members of the extended family, perhaps grandparents, children, and grandchildren, might live together in the same household. Extended families might live in the same community. An extended family might live a long distance apart in a large country like Canada, or live in other countries. Maintaining ties with your extended family can provide you with a sense of your personal history and your culture.

Couples often consider themselves to be a family. The members of a couple are called spouses. Sometimes a *spouse* or partner is also referred to as the "significant other." Most nuclear families were childless couples before they had their first child, but some couples might remain childless. The largest group of childless couples in Canada is the empty-nest families, older parents of nuclear families whose children have left home. Most spouses are married, but an increasing number are cohabiting. The relationship with their significant other is still considered to be their most important relationship by adult Canadians.

There are many variations of these basic family forms. In some families, the two parents are cohabiting in a common-law marriage, where a couple is formed without legalized marriage. Children in an *adoptive family* might be of a different race than their parents or their siblings. Children sometimes live with a *foster family* for temporary care until they can return to their biological family. Some children live in families in which the two parents are a cohabiting couple of the same sex. A child might live with grandparents instead of parents, or be raised by older siblings or other extended family members. Who the members of your family are is one way in which your family is unique.

Your extended family provides you with a group of people that care about you and with whom you have common traits. They are often your closest relationships and give you a sense of place and belonging.

CHECK Points

✓ Identify and describe the different types of parent–child families.

✓ What are the different types of couples?

✓ Identify examples of each family form in popular television shows.

✓ How have nuclear families changed since the 1950s?

Shannon and Nahid's Story

"You have it so easy! Your grandmother makes dinner and even makes you a snack after school. You do not have to go home every day and start supper." Shannon turned to Nahid beside her on the steps. This was not the first time the friends had compared their families.

"My grandmother does make good samosas, but she's always there! Every day she asks me what I have done in school and reminds me to do my homework. She makes me feel stupid sometimes."

"Nahid, she is so nice. I would love to have my grandmother live with us, but she is in Vancouver so I do not see her very often. It is just me, mom, and dad."

Nahid shook her head. *"A small family is better. I would like to be able to talk to my mother the way you talk to your mom. I feel surrounded, with three older brothers telling me what to do as well as my parents. And I am not allowed to talk to boys."*

"Well that would be a problem. Did you see Jay today? He is so cool! But I feel lonely when mom works late. She does not have time to talk to me then, and I have to cook for me and dad. I wish I had a brother or a sister to do things with, and to share the dishes with.

Oh, there's the bell. See you later at your locker."

What do you think?

1. Identify Shannon's and Nahid's family forms.
2. What advantages do they identify for each form of family?
3. Which family is most like your own?
4. Which family would you prefer to live with? Why?

Reflections and Connections

1. Who are your family members and what is your family form?
2. Have you always lived in the same form of family?
3. Should people be married to be considered spouses? Why or why not?
4. Do you think children are necessary to form a family?
5. What do you think would be the ideal family form for Canadians at the beginning of the twenty-first century?

The Seasons of Family Life

Most people live in two generations of family over a lifetime. The family you were born or adopted into is called your family of origin. This family was responsible for nurturing and socializing you, and will soon launch you into the adult world. As an adult, you may meet someone whom you will spend the rest of your life with. When you, in this couple, have your own children, you will form a family of procreation. These two families differ according to the role you play in the central parent–child relationship. In your family of origin, you are the child; however, in a family of procreation, you would be the parent.

Families grow and develop as they meet the challenges they experience. The responsibilities of family life begin when a couple forms. They increase tremendously if they have a child, and continue to change as that child grows up. So the cycle of family life begins. Each family's life cycle is unique, depending on the form of the family and the timing of the additions and changes to the family. Some families stay together, but some parents divorce. Unfortunately, some family members die due to illness or accidents. Families survive through meeting the challenges presented to them and making adjustments, often with the support of the extended family, friends, or their community.

Like individuals, families also have developmental tasks. Couples must create a relationship that enables them to share their lives and their responsibilities in a way that suits them as individuals. They must adjust this relationship to meet the demands of parenthood. Couples will undergo adjustments again when the children leave home. Parent–child relationships must be established when a child is born. As the child grows and develops, becoming an adolescent, then eventually an adult, the parent–child relationship must change. Individuals continue to develop into old age. Families must grieve the loss of those who die and adapt to life without them. Families at the same stage in family life share common interests and challenges that can draw them together for support and friendship.

Every family has the same functions and responsibilities, yet each family is unique. Though their family forms may look similar, one might be a nuclear family with adopted children, while the other may be a blended family.

Families at the same stage of the family life cycle have common interests because they face similar challenges.

Families with adolescents will often find that all members of the family are facing similar developmental tasks at the same time, despite the differences in their ages. You have learned that the major task of adolescence is forming an identity in order to prepare for a meaningful adult life. As a teenager, you expect to have opportunities in the future for achieving your goals. At the same time, adults in your family may be reviewing their lives to see if they've achieved all their goals. Teenagers and their parents often share a similar need to make decisions about individual goals, relationships, family, and careers. You and your parents will have different perspectives on how these important issues can be dealt with.

Check Points

✓ Why do families change as time goes by?

✓ What are the similarities between the developmental tasks of teenagers and their parents?

✓ Why might teenagers and their parents have a different perspective on life?

Reflections and Connections

1. At which stage of family life is your family?
2. What new developmental task has your family faced at this stage?
3. What was the major challenge for your family five years ago?
4. What will be the challenges for your family five years from now?

The Importance of the Family

Your family was your first environment; therefore, you will compare your experiences through life with the early experiences you had with your family. If these years were happy ones, you will treasure the memories and expect future happiness. If you feel that you have had unhappy experiences, you might be concerned about the future. Evaluate your childhood and realize that you can change your destiny with some definite choices and a positive outlook.

Your family is an important asset as you set out on your journey through life. So important are human families that there has seldom been a society that was not organized into family groups. Families serve a clear purpose in society—to carry out the functions that are the primary responsibility of the family. No other part of society holds sole responsibility for this role. Because the family is the basic unit in all societies, it is easy to take families for granted. Families have changed and will probably continue to change in the future, but one thing is clear—families are here to stay.

Society as we know it would be simply unimaginable without families.

Benjamin Schlesinger

Summary

1. Defining the family determines the rights and responsibilities of individuals.
2. A family is defined as any group of two or more persons, who are bound together over time by ties of blood or mutual consent, birth, and/or adoption and who, together, assume responsibility for the functions of families.
3. Families are responsible for the functions of families, including providing physical care, adding new members, economic support, raising and socialization of children, meeting emotional needs, and social control of family members.
4. Family forms vary in the nature of the relationship between the adults, whether or not there are children, and the number of generations.
5. Families are challenged by new developmental tasks at each stage of the family life cycle.
6. Adolescents and their parents face similar developmental tasks.

In this chapter, you have learned to

- explain how family responsibilities change through the family life cycle;
- summarize the universal functions of families and their effects;
- compare several definitions of the family; and
- describe the variations in family forms and relationships.

Activities to Demonstrate Your Learning

1. Create a collage of images and words representing the diversity of families in Canada today. Include images and definitions from a variety of sources to represent both realistic and stereotypical points of view. Organize the display to distinguish between the stereotypes and reality.

2. Using social science methods, investigate how families in your school community fulfil the functions of the family. Formulate a research question, design and conduct a survey, and graph the results using spreadsheet software. Form a conclusion and write a brief report.

3. Using a chart, outline the costs and benefits of different types of families. Consider factors such as love and affection, space and privacy, supervision and responsibility.

4. Using social science methods, investigate how the challenges and benefits of family life change from stage to stage. Formulate a research question, and design and conduct an interview with a grandparent or an older adult of grandparent age. Form a conclusion and write a report using quotations from your interview.

5. Create a storyboard for a commercial representing the diversity and strengths of families in Canada.

Enrichment Opportunities

1. What do you think families will be like 10 years from now? Write a newspaper article in which you outline the benefits of this form of family.

2. Write a short story about life in a future society without families. In your story, explain how the family functions are performed in the absence of families.

3. Work collaboratively with a group to prepare a scene that would be a pilot for a new television sitcom. In your program, you propose to present a realistic image of family in Canada to counter the stereotypes.

Families in a Social Context

Human beings have lived in families since before recorded history, but families today are different from those in the past and from one country to another. How have families changed throughout history? Why do families change? How do families in Canada differ from those in other countries? What are the influences on families in Canada today? What would your life be like if you lived in a family in a different time or in a different place?

By the end of this chapter, you will be able to

☀ compare current lifestyles with those in the past and in various cultures;

☀ compare the roles of adolescents within Canada and in other countries;

☀ compare the roles of males and females within Canada and in other countries; and

☀ analyse the impact of social and political policies on lifestyles.

You will also use these important terms:

age of majority	patriarchal
clans	polygamous
consumer	public services
divorce	social policy
governments	taxation
monogamous	

The Evolution of the Family

There has seldom been a society that was not organized into family groups. Attempts to develop other ways of organizing society to carry out functions that are the responsibility of the family have not survived very long.

Anthropologists study human civilizations to learn about human culture—why people do what they do. There is no written history of the earliest human societies. Anthropologists study isolated human societies who do not seem to have developed technology or have been influenced by other societies for clues about human societies in the past. Many theories have developed to try to explain the development of human civilization. At the centre of human civilization is the family.

The human family might have developed because of one unique characteristic of human beings—we have large brains. Large brains enable us to use language, to think and discuss ideas, to develop technology, and to feel emotions. So, human infants are born with large heads to hold their large brains. Human babies have the potential to be the most intelligent of all animals, but they are born helpless.

Babies need extensive care and socialization for at least five years before they can begin to become independent. Until recently, they needed to be fed by their mothers for at least a year. Once humans understood sexual reproduction, men probably wanted to make sure that their own children survived. Perhaps the earliest humans formed families to make it possible for a mother to look after a child. A man who stayed with a woman for years could defend her and their child, and help provide the necessary food. The earliest families were probably based on mutual support for survival.

Families differ in how they meet their needs, and how they divide the family work. How a society organizes and administers the use of resources to meet their needs is called economics. As the economics of human society changed, families also changed. In larger human societies, governments and judicial systems established rules concerning family life. Organized religions also set expectations for all areas of life, including the family. Technology changed how work was done and, later, who did the work. As we enter the twenty-first century, families continue to change.

Hunter-Gatherer Families

The earliest families were hunter-gatherers who depended on gathering plant foods and hunting to feed themselves. Finding food was a full-time occupation for both men and women. Women were responsible for nurturing young children as well as gathering fruit and vegetable foods every day. Men left the family to hunt

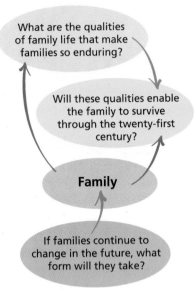

What are the qualities of family life that make families so enduring?

Will these qualities enable the family to survive through the twenty-first century?

Family

If families continue to change in the future, what form will they take?

CHECK Points

✓ What are the characteristics that make humans different from other species?

✓ Why do human babies need to be cared for so much longer than babies of other animals?

✓ What is one theory about the origin of human families?

because they were not able to nurse children. Before they had weapons, hunters had to pursue a large animal for days until it became tired and could be caught and killed, so meat was eaten only occasionally. One hunter could not provide meat for many people, so families probably consisted of one man and only one woman and their children. Nomadic groups of several couples and their children travelled together in search of food.

The earliest Canadians, the Aboriginal peoples, were mostly hunter-gatherers. They lived in groups of families and travelled seasonally to find food and to follow the animals that they hunted. The division of labour was based on age and sex. Females cared for the children, gathered food, prepared the food, and made clothing and other domestic objects. Males hunted and made tools for hunting. Males also defended their hunting territory against neighbouring tribes. Because of the conflict between hunting-gathering societies and agricultural societies for use of land, the few hunter-gatherers left in the world today are in remote and isolated places, such as the Amazon rain forest of Brazil.

Early hunters would leave their families for several days at a time.

The Amazon River Indians

The Urueu-Wau-Wau Indians in the rain forest of Rond'nia in the Amazon valley are one of several tribes who still live as hunter-gatherers. They live in a territory that was declared off-limits to outsiders by the Brazilian government in order to protect their way of life. Members of this tribe wear no clothes but use haircuts, tattoos, and make-up to decorate their bodies. They live in large woven straw houses beside several others around a central meeting space. They use the plants of the rain forest for food, for medicines, and to make poisoned arrows for hunting. Their lifestyle is suited to the environment of the rain forest.

Should the few hunter-gatherer societies left in the world be protected from modern society?

Amongst the Urueu-Wau-Wau, the roles of men, women, and children are clearly defined. They were taught by their elders how to perform their roles as men or women. Women gather plant food, prepare the food, maintain the home, and care for the children of the village. Men go off to hunt for days at a time. The whole village celebrates their return from the hunt with a ritual dance and a communal feast. Adolescent boys contribute by making tools and hunting. Adolescent girls contribute by gathering and preparing food and caring for children. At the end of the twentieth century, the Urueu-Wau-Wau lifestyle had become a living museum of the hunter-gatherer existence, protected by the government from the influence of drug companies who want to learn their technology of using plants for medicines.

career Link

What is an archeologist?

An archeologist is a person trained in the study of people, customs, and life of ancient times.

What do archeologists do?

Archeologists study, analyse, and conduct research. They often go to sites where artifacts of ancient civilizations have been discovered and conduct "digs." While digging, they discover important information about how people lived hundreds or even thousands of years ago. By finding, examining, analysing, and cataloguing these artifacts, they provide historical evidence about families and family life.

Where do archeologists work?

Although their work often requires them to travel throughout the world, archeologists often work for government services such as museums and historical societies. Some work is also available in the educational and business sectors.

CHECK Points

✓ Why would finding food be a full-time job for hunter-gatherers?

✓ Why is there no flexibility in the roles of males and females in hunter-gatherer families?

✓ What did adolescents do in hunter-gatherer families?

Reflections and Connections

1. There is no record of the earliest families. Do you think they were *monogamous*? At what age do you think they "married?"

2. Why are there so few nomadic people in the world today?

3. The hunter-gatherers of Canada did not survive exposure to Europeans. What happened to their lifestyle?

4. What are the benefits of the Urueu-Wau-Wau lifestyle?

5. What changes might occur as the Urueu-Wau-Wau begin to interact with the modern world?

Agricultural Families

The development of agriculture changed the pattern of human families. When humans were able to grow plants for food and domesticate animals, families were able to settle in one location and build a permanent home. Agriculture could provide much more food, but it required a lot of manual labour. Early agricultural families were large because more people to work meant the family could acquire more land and become wealthier. Since one man could afford to feed several wives in order to have more children, *polygamous* families became common.

Women often married young to produce more children, but men waited until they were older. One woman could care for many children, so some women could also work in the fields. Because a family needed land for agriculture, young adults brought their spouses to live with their families on the family farm, thus forming extended families. As families expanded and acquired more land, families grew to become *clans* of many related extended families living in the same area.

Farming in Fiji

Agricultural societies are common in most parts of the developing world. In Fiji, sugar cane is the major crop. Families rent their land from the government. Amongst Indo-Fijians, the extended family ensures that there are enough people to work the land and provide support for family members. Older generations of men and women work in the fields and care for the animals. Traditionally, the oldest son would have worked on the farm alongside his father, but technology has reduced the amount of labour required. Nowadays, as is the case in most countries, the family depends on the son's employment in town to supplement their farming income so that they can buy equipment.

In the extended Indo-Fijian family, roles in the family are assigned according to age and gender in the traditional Indian manner. The oldest male is the head of the household. His wife looks forward to an arranged marriage for her son so that he brings a daughter-in-law into the home. It is the role of the daughter-in-law to look after the home, prepare the meals, do the laundry, and care for the children. Her task is also easier, now that families can buy electric stoves, refrigerators, and washing machines. Because marriages are arranged, husbands and wives learn to like each other and share their lives, but love is a bonus, not an expectation. Families encourage sons and daughters to get an education and family planning programs enable families to have fewer children.

Men and women can work side by side in agriculture to produce food for the family.

CHECK Points

✓ What would be some advantages of an agricultural lifestyle?

✓ Why were extended families more common among agricultural families?

✓ What are the benefits of having many children?

Reflections and Connections

1. What were some advantages of polygamous families?
2. Why do you think polygamous families are so uncommon in the world today?
3. How do farming families in Canada compare with the family in Fiji?
4. Why do many farming families need employment off the farm to supplement their family income?

Pre-Industrial Families

The development of technology over the last thousand years enabled some families to work at crafts as carpenters or blacksmiths, to produce the things required by other families. These pre-industrial families settled in villages and early towns. The skilled craftsman would work in the family home where women and children could help with the work. These families often included apprentices, young people from other families who were learning the craft, and servants who would do the household work. The craftsman could not support many children since only one or two sons could continue his business. Families were monogamous and had fewer children. Definitions of family at this time include the father as head of a household, his wife and children, servants, and apprentices.

The settlers who came to Canada brought pre-industrial society with them. Military and government officers and some craftspeople congregated in villages and towns, but the majority of settlers spread out in self-sufficient homes across the country. Small fishing villages, home to several fishing families, dotted the coastline of the Maritimes. Distinctive seigneurial settlements developed along the shores of the rivers in Quebec. In Ontario, families lived on isolated homesteads, and in the Prairies, families lived on vast farmsteads.

The European settlers who came to Canada brought the European tradition of monogamous marriage with them. Men and women worked side by side to clear and farm the land. Women cared for children and maintained the household. Children joined in the work as soon as they were able to carry things. Although there was a division of labour based on sex, there was no end to the labour. Life was hard and the work was endless for men, women, and children. Older children and adolescents were often sent to work for other families. Boys would work on a farm or learn a trade as an apprentice, and girls would do household work as domestic servants. Young people married later in order to be able to afford a farm, or a shop, and a home.

Early settlers worked side by side to clear the land, build a home, and produce food for their family.

CHECK Points

✓ How would craftsmen and their families obtain the food they needed?
✓ Why did the lifestyles of men and women differ in the pre-industrial family?
✓ Why were pre-industrial families smaller?
✓ What did apprentices and domestic servants do?

Reflections and Connections

1. What crafts were needed in early Canadian settlements?
2. What would be your lifestyle if you were a teenager in the pre-industrial family?
3. Why were apprentices and servants included as part of the family?
4. Why would the families of craftsmen be wealthier than the families of farmers?

Urban Industrial Families

The development of machines which could do the work of many people caused an industrial revolution over the last two hundred years. Because people left home each day to go to work in factories located in towns and cities, paid work and household work became separated. Men, adolescent boys, and unmarried girls went to work. Women with young children stayed home with sole responsibility for unpaid household work until the children could be left unattended or with older children. About one hundred years ago in Europe and North America, child labour laws banned children from the workplace and sent them to school. Because children were no longer able to earn income for the family, parents had fewer children and families became smaller. Although many working-class women worked in the factories, society considered the breadwinner family, consisting of a man, his stay-at-home wife, and his children, to be the ideal family.

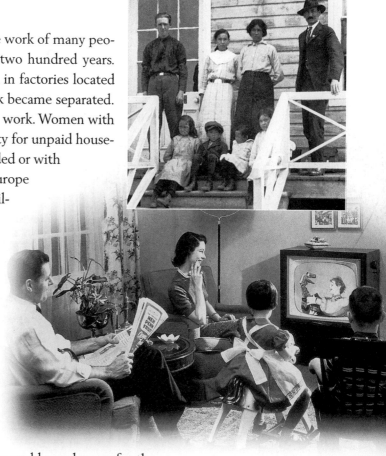

By the 1920s, the breadwinner family became a *consumer* family. The husband was the earner and the head of the household. He was the link between the family and the larger society. The woman's role was to be a wife, mother, and housekeeper for the family. She was also the family's major consumer of mass-produced clothing, furniture, and food products. Children attended school and played, protected from the hard work of the adult world. By the 1950s, children were staying in school until their mid-teens, and teenagers became a distinct age group between children and adults. By the 1970s, motivated by the desire to buy more goods and the availability of jobs, married women started to work outside the home. Children who were not needed for farm work might be sent to earn their keep as a servant or as an apprentice to a city family.

Top: The breadwinner family became the ideal family for Canadians in the early twentieth century.

Bottom: The consumer family enjoyed the leisure time offered by an industrial society.

The Urban Korean Family

In modern Korea, people take pride in the improvements they have made in their lifestyles since the end of the Korean War. Homes are traditional in appearance, built inside a courtyard that suggests the privacy of family life. There is a family car parked outside the house, and inside it is clear that Koreans now live a consumer lifestyle. Families sit together, cross-legged on the floor, to enjoy spicy Korean meals that have been prepared in modern kitchens. Families spend time watching television together before the children go to their rooms to do their schoolwork. Traditional values of knowledge and quiet family life are still held in a modern urban lifestyle.

CHECK Points

✓ Why did men begin to work away from the home after the industrial revolution?
✓ How did the role of housewife develop?
✓ What made teenagers a distinct group, different from children and from adults?
✓ Why were families able to buy more goods by the 1920s?
✓ Why did many women begin to work outside the home by the 1970s?

Families work hard to achieve the comfortable urban lifestyle. Men work long hours, up to 12 hours a day, to earn the required income. In the extended family, older relatives help to do the housework, so women are encouraged to do volunteer work to contribute to the community. Children are expected to work long hours at their schoolwork. Tutoring is available at the school after the regular lessons to help children compete for the university education they need to get respected jobs. Hard work pays off with a lifestyle that includes the conveniences of modern life.

Reflections and Connections

1. Korean mothers are encouraged to do volunteer work. What are women encouraged to do in Canada?
2. How is the role of teenager different from that of child in a modern urban society?
3. Is it possible to maintain traditional lifestyles in a modern consumer society?
4. Why do you think education is so important to modern urban families?

Control of the Family

The Role of Law

Governments develop and enforce laws that regulate behaviour within a society. Family law defines the relationships among people to ensure that individuals do not become a burden on the society. By setting a minimum age for marriage, a society ensures that married couples can be economically independent. In Quebec in the 1700s, a man needed the permission of his father to marry until he was 25 years old, the *age of majority*. Now that the age of majority across Canada is 18 years of age, a teenager can marry at 18 without his or her parents' permission, or as young as 16 years of age in some parts of Canada with the permission of both parents. This seldom happens. Because students must stay in school until the age of 16, and most students finish Grade 12, they cannot afford to get married. The fact that the law allows something does not mean that it happens regularly.

The Married Woman's Property Act was passed in Canada in 1870. This law permitted a married woman to keep her own name and legal identity, after marriage. It also allowed her to control the property that she owned before marrying. In spite of this law, women were financially dependent on their husbands because very few had property of their own. Later, family law reforms introduced a concept called "community property" which means that a couple must divide the property they acquired together during marriage equally if they get divorced. These and other legal changes reduced a woman's economic dependence on her husband. Today many women earn an income, so Canadian marriages have became more equal partnerships.

Family law has changed to reflect new patterns of family life. The divorce law was changed in 1968 to allow **divorce** when a couple has irreconcilable differences. This reflects the change in marriage that occurred in Canada in the twentieth century. When couples married for economic reasons and men had authority over their wives, women rarely left their husbands, regardless of their differences. However, women who earn their own incomes do not need a man to support them. Today, most men and women have the option of marrying for love and companionship, not economic reasons. The higher divorce rate in Canada might have resulted from the expectation that we will be friends with our spouse and have the option of separating if we no longer get along.

The Role of Social Policy

Governments also determine **social policy** by collecting money through **taxation** to provide services to help families perform their functions. Income taxes are collected as a portion of an individual's income. Taxes are also collected when you purchase goods and services. Money raised through taxes may be used to pay for **public services** such as health care, education, and fire and police protection for everyone in the society. Tax revenues may also be used for transfer payments to supplement family income for the poor, the elderly, the disabled, and the unemployed. Where government assistance is available, smaller nuclear family units can support themselves without depending on the extended family or on numerous sons and daughters. Government support replaces the economic role of the extended family.

Very few 16-year-olds are married in Canada today.

The Role of Religion

The earliest Canadians, Aboriginal peoples, believe that everything in the world is connected—humans, animals, plants, and nature. This belief focuses on the natural cycles of time, of the seasons, and of birth, growth, and death. Early Aboriginal family life enabled people to live out the natural cycle of life in harmony with one another and with nature, so families were the centre of Aboriginal spirituality and society. Children were

Traditionally, children were treated with respect by Aboriginal parents.

Hutterite families live a simple communal lifestyle.

considered a gift from the spirits. Europeans were surprised by the respect with which Aboriginal children were treated. They were expected to learn by observing and listening to stories from the elders rather than by the harsh discipline that was common in Europe. Families were quite stable because family membership determined the role an individual played in the community.

The Roman Catholic Church had clear expectations concerning marriage and family life for the French and Irish settlers in Quebec. Catholics believed that the purpose of marriage is procreation, having children. Using contraception was considered a sin, so families had many children. In early Quebec, families were *patriarchal* families, in which the wife and children are expected to obey the father. Divorce was not allowed, but it was also not practical if there were many children to support and women were not supposed to work outside the home. Young Catholics learned to carry out their obligations to their families.

Many families that came to Canada in the early 1900s left Europe because they were being persecuted for their religious beliefs. When they settled in Canada, they looked for farmland to establish their own close-knit communities based on their beliefs. In Hutterite communities, for example, families own property communally. Although families live in family homes, men and women live very separate lives. Men work together at farming and maintaining the equipment for the community, and women work together at producing food, clothing, and other essentials. All families eat together in a communal dining hall, with men, women, and children in separate areas. Children spend their days in a nursery until they reach school age. Adolescents leave school early, and marry young. Hutterites choose to live a simpler agricultural lifestyle of the nineteenth century, and to live communally, because of their religious customs.

The Role of Culture

By the end of the twentieth century, most of the world's religions were represented in Canada. Today, many different cultural beliefs about family life and the roles of family members exist. For example, many traditional Hindu and Muslim families believe in arranging marriages for their children. Arranged marriages are based on a careful consideration of social, cultural, and economic factors to determine whether the two young people are likely to be compatible. Nowadays, when the parents think there is a good match, the young man and woman are given an opportunity to meet and decide whether they like each other and want to marry. Many young Hindu and Muslim men and women accept that the wisdom of experience might be more successful than romantic attraction, but others want to adopt the custom of love-marriages. Cultural beliefs about the roles of men and women, parents and children, and the family vary widely in Canada today.

Being Brought Together:
The Practice of Arranging Marriages Continues to Thrive

By Rashida Dhooma

Even as a teenager, Tiger Ali Singh knew his parents would choose a wife for him. Today, four years after entering into an arranged marriage, the 27-year-old WWF wrestler says he wouldn't have it any other way. "Who better to find me a life partner than my parents, for whom I have the utmost respect and regard as my role models," says the man who received marriage proposals from around the world when word spread that his parents were looking for a wife for him.

His trust in their ability to find him a suitable bride paid off when they introduced him to Harmeet Kler of Singapore, he says. "I couldn't have chosen better myself." Singh and Kler have a 2-year-old son, Gurjeevan, with another baby expected in March.

In Singh's case, the community networking machinery helped find him his mate. But because the network here isn't as strong as in India and other countries where arranged marriages are the norm, increasingly families are relying on personal ads in newspapers and on the Internet.

Each week, English-language South Asian newspapers carry hundreds of ads of mainly professionals seeking marriage partners with similar qualifications.

"The success rate of these ads is very high," says Ajit Jain, associate editor of India Abroad, which carries about 200 ads in its Canadian edition. Unfortunately, he says, arranged marriages get a bad rap when extreme cases of mostly young girls being forced to marry older men make headlines. Two weeks ago, an Afghani woman was beaten, bound, and gagged for refusing to marry a relative in an arranged wedding.

"There's no denying that forced marriages happen here to a certain extent, but in most cases both parties have the right to say no," says Ushi Choudhry, professor of nursing at Seneca College, who's "fascinated by the West's fascination" with this subject. The practice, she says, continues not only in South Asian cultures, but among western cultures and royalty, the wealthy, and other groups to protect themselves, their money, or property.

Arranged marriages among South Asians, Choudhry adds, is a union of two families, and is reflective of their thinking that it's for the collective good, as opposed to an individual's needs.

"That's why families are closely involved in evaluating a person entering their families. They don't want someone who'll upset the status quo," she says. "Marrying someone of your own background ensures an easier transition and there's little, or no confusion."

She does, however, concede that these unions place pressure on couples, especially women, because of negative implications if marriages fail. "Whole families sink," she says., "and women are usually the ones who suffer most because they're on the outs if they leave." Children rely on their elders for guidance because "experience knows best," Choudhry says. "We've great respect and reverence for elders, and we value their lived experience over the so-called educational experience. Life is the true teacher," she says. These parents, Choudhry adds, have their children's interests at heart.

But with children growing up dominated by western influences and "in a more reflective way," there's the inevitable conflict as children challenge them. "These children think their parents aren't in touch with reality," says Choudhry, who calls for compromise on both sides. Children, she says, should ask "is there any merit" to their parents views, and the older generation has to come to terms with the changing world around them. Arranged marriages, she says, are successful to about the same degree as ones for love. "When I came to North America about 30 years ago, I naively wondered why the divorce rate was so high if everybody was marrying for love, " she says, laughing.

Neera Malhan, 26, who was introduced to her husband Rajiv six years ago, says coming from the same religious and cultural backgrounds has helped their marriage. "We even live with Rajiv's parents, and it's not a difficult situation," says Malhan, a lab technician who's expecting a second child next week. Their son, Jai, is 18 months old. Rajiv, a market research analyst at a Mississauga cardiology clinic, regards his as a semi-arranged marriage because he and Neera took matters into their own hands once they were introduced. "Either of us could've said no," he says.

The Toronto Sun 01-11-99
Reprinted with permission of Sun Media Corp.

What do you think?

1. What arguments are given in support of arranged marriages?

2. What are the arguments against arranged marriages?

3. Can people fall in love in an arranged marriage?

4. How does the rate of divorce compare for arranged marriages and love-marriages?

5. How do you think an arranged marriage would differ from a love-marriage after the couple are married?

6. If your parents were going to arrange a marriage for you, what kind of person would they choose for you?

Technology and Family Life

Do cell phones and pagers improve your relationships?

At the beginning of the twenty-first century, your lifestyle resembles the science fiction stories that your parents read when they were in school. You can talk to people using small devices that fit into a small pocket. You can communicate with others around the world, or order a pizza for delivery at the keyboard of your computer. It is possible to live your life, working, shopping, and visiting friends, without leaving your home. Surgeons can transplant organs from donors to another person. Scientists have cloned exact replicas of animals from samples of their tissue. Astronauts are building a larger space station for the men and women who live and work in space for months at a time. There seems to be no limit to the technology that we can develop.

Back at the beginning of the twentieth century, electricity became available in Canadian homes. Perhaps the most significant technological development was electric light that enabled us to live and work beyond sunset. It allows us to set our own patterns of sleeping and waking. The number of hours spent preparing meals, doing family laundry, and cleaning the home was reduced by the introduction of stoves, refrigerators, washing machines, and vacuum cleaners. Electricity has increased what we are able to do so that the pace of our lives is much faster than it was a century ago. Having the technology to do our work requires that we work to pay for it.

The development of communications technology changed how we communicate. The telephone, which entered our homes at the beginning of the twentieth century, and the cell phone which became available at the end of the century, allow us to maintain contact more easily when we are away from our family and friends. The inventions of radio, record players, and, later, television brought music, theatre, and world news into our homes, and enabled families to enjoy this entertainment together. Computers enable us to communicate instantly around the world and bring information to our

homes. The technology that allows us to be active at night in sub-zero temperatures also changes the working conditions of Canadians in ways that can be difficult for families. Sometimes the use of this technology tempts people to spend so much time being entertained that they do not find enough time for schoolwork, physical activity, family, or friends.

The Contemporary Family

Today, the family is the focus of a set of influences. Governments make laws that regulate the formation of families and the economic relationships between people. Religion specifies the expectations of people towards each other. Education affects the role of children and adolescents in the family and their future social status. Medical technology, which reduced child mortality and increased life expectancy, has changed the nature of families. Businesses affect the spending habits and the division of labour within families. The media presents you with images of family life. Your role as a man or woman, as a child, or as an adolescent within your family is affected by the many agents existing within your society.

Reflections and Connections

1. Describe the impact of the most recent technological development to affect your family lifestyle.
2. In what ways has technology had a negative effect on Canadian lifestyles?
3. How does communication technology affect how you communicate with people?

So many influences affect the family. Yet, you will have many options to select from if you choose to form a family of your own some day. You and your partner will create a unique family based on your choices about how old you are when you get together, whether you marry, the roles you will each play, whether you have children, and how you will raise them. You can decide what type of home you want, who will share your household, who will work outside the home, and so on. What will your future family be like?

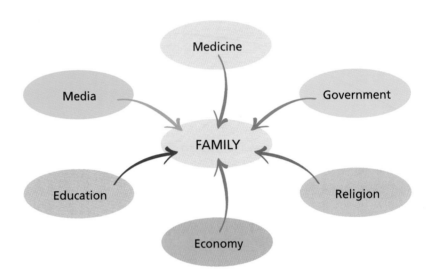

Summary

1. Humans probably formed families because their babies are dependent for longer than other animal babies.

2. The earliest families were nomadic hunter-gatherers consisting of groups of man–woman families and their children.

3. The settled lifestyle of agricultural families allowed them to develop technology and support larger families.

4. Pre-industrial families carried out trades at home, and included servants and apprentices as part of the family.

5. The industrial revolution resulted in a separation of work and home, and the smaller breadwinner family.

6. All of the historical forms of family continue to exist in parts of the world today.

7. Families are affected by laws, social policy, religion, and cultural tradition.

8. The modern urban family is a smaller nuclear family that is supported in its functions by government services.

9. Technology has improved the standard of living for families, but also created a media culture which consumes much of our time.

In this chapter, you have learned to

- compare lifestyles now and in the past and in various cultures;
- compare the roles of adolescents within Canada and in other countries;
- compare the roles of males and females within Canada and in other countries; and
- analyse the impact of social and political policies on lifestyles.

Activities to Demonstrate Your Learning *Either 1 or 2*

1. Using the Internet, investigate the lifestyle of the Amazon Indians. Determine how they are being affected by contact with the modern world and write a brief update.

2. Using the resources of your library, compile fact sheets describing the lifestyle of adolescents in other cultures to share with your classmates. Identify the criteria that can be used and compare lifestyles to identify similarities and differences. Record your conclusions about the role of teenagers around the world.

3. Investigate the food resources available in another culture and prepare a typical dish. Explain how the foods would have been produced, sold, prepared, and eaten in the traditional culture.

4. Using social science research methods, conduct interviews with adult immigrants to Canada to answer the research question: How do the roles of men and women vary in other countries? Compile the results in a chart. Write a brief report.

Enrichment Opportunities

1. Read an historical novel with an adolescent protagonist. Compare your lifestyle with that of the character in the novel.

2. View a movie about family life in another culture. Compare the life of your family with the family life depicted in the movie. Evaluate whether the depiction of the family is accurate or stereotypical.

A Portrait of the Canadian Family

Do your parents tell you about what life was like when they were your age? Do they talk about life back in their home country? Do your grandparents reminisce about the way things used to be—even talk about "the good old days?"

Welcome to the club. The state of the family is one of the most common topics of conversation with opinions ranging from "the family is alive and well" to "the institution of the family is dying." Your family story is your personal history, but the future of the family is also your future.

By the end of this chapter, you will be able to

* compare the roles of family members in a variety of family forms;

* distinguish between facts and opinions about the current state of the Canadian family;

* summarize the impact of social, economic, legal, technological, and environmental changes on the lifestyles of individuals and families;

* analyse the impact of social and political policies on the lifestyles of individuals and families; and

* explain the effect of diversity in family form on specific aspects of family lifestyle.

You will also use these important terms:

conjugal arrangement	life expectancy
custody	standard of living
demographic trend	statistician
infertile	transfer payments
irreconcilable differences	

The Current State of the Canadian Family

Which side are you on? Is the institution of the family dying—outdated and no longer able to meet the needs of Canadians? Or do you believe that the family is alive and well—still the best way of meeting the needs of people, young and old? These are the opposing points of view on the current state of the family in Canada that have been argued for over 50 years, since social scientists first began a serious study of families in Canada. As with any controversy, both sides have strong arguments and present evidence to support their point of view. What evidence should you consider in order to form your own opinion?

Your opinions about the family are based on your own family experience, on what you hear people say, and on the information you gather from the media. The family is a very common topic of discussion. For example, the increasing number of divorces is blamed for many of the problems of our society. On the other hand, people point to the high rate of remarriage as a sign that people do want to be married. The media presents the news with an emphasis on the bad news: the cliché today is becoming, "no news is good news!" On television programs, conflict is necessary for an interesting story but the story may present an unrealistic picture of family life.

There are more than 7 800 000 families in Canada. In 1996, the last year for which Statistics Canada numbers are available, 80 percent of Canadians were living in families. By far the majority of families include "married" couples.

**Families in Canada in 1996
With or Without Children**

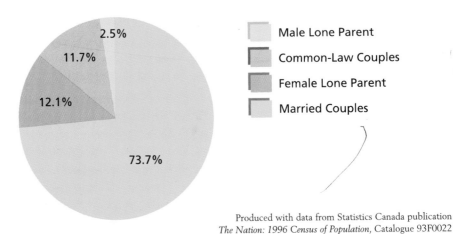

Produced with data from Statistics Canada publication
The Nation: 1996 Census of Population, Catalogue 93F0022

Although the number of marriages is dropping, the number of couples in common-law relationships is growing rapidly in Canada, especially in Quebec. This creates the question—why were 20 percent of the people not living in families?

The Single Life

One reason for the number of people who do not live in families is that young adults seem to be moving out independently and waiting until their late twenties to marry. Why are you likely to get married later? One reason could be that employers today expect new employees to be better educated. To prepare yourself for the competitive job market, you might be older when you finish your education. Low employment opportunities for young adults without post-secondary education or apprenticeship training might mean that you will change jobs frequently in order to move into a job that fits your career goals. Low-paying entry-level jobs combined with student loan debts might mean that you live with your parents longer in order to save money. Achieving career goals can also mean working long hours which can limit your opportunities for meeting that special someone you want to marry. For many young Canadians, marriage must wait until careers are underway.

Many people today want to enjoy each other's company without making a commitment to live together forever. Many Canadians today say that they believe that sexual relationships outside of marriage are acceptable. Some people have intimate sexual relationships and delay marriage until they are ready to have children. Others say they enjoy the freedom of living alone and calling up friends when they want to share an activity. Some people stayed single in the past, but today's society provides more social opportunities outside of the family. The single lifestyle is more visible and acceptable today. Being single is a lifestyle that some individuals choose for a variety of reasons.

Top: Being single does not have the same negative image that it had in the past.

Bottom: Dating can present challenges to the newly single adult in search of a life partner.

CHECK ✓Points

- ✓ What are some opposing opinions on the family in Canada?
- ✓ What are some reasons why the percentage of people not living in families is increasing in Canada?
- ✓ What are some theories to explain why Canadians are delaying marriage?
- ✓ Why is being single more acceptable today for both men and women?
- ✓ Explain the increase in number of older single people.

Another group of people who are not living in families are older people who are widowed or divorced, who have no children, or whose children no longer live at home. Some choose to live alone because they feel that no one can replace a beloved spouse who has died. Others choose to live alone because of a bad experience in a marriage that ended in divorce. Many widowed or divorced adults, however, are dating. They are hoping to find a new partner who will provide affection and companionship. Older people, and older singles, are more numerous because of our increased life expectancy.

Conjugal Arrangements Experienced By Women, By Age

| | Age in 1995 | | | | |
	60-69 %	50-59 %	40-49 %	30-39 %	20-29 %
Proportion of all women experiencing:					
At least one union	96	97	96	94	87
At least one marriage	96	95	92	84	66
First union starts with marriage	95	91	78	56	35
At least one common-law union	8	22	35	49	59
First union starts with common-law	1	6	18	38	52
At least one separation	25	32	40	43	—
At least two unions	14	27	34	39	—
At least two separations	8	13	16	—	—

— amount too small to be expressed

What do you think?

- A *demographic trend* is a pattern of change in the behaviour of a population. A trend does not predict the behaviour of one person but it can suggest the likelihood of the person behaving in a certain way.
- A *conjugal arrangement* is a live-in sexual relationship. Examine the statistics and suggest some trends in conjugal arrangements for Canadian women.

The Daily, Statistics Canada, Thursday, March 16, 2000.
Canadian Social Trends: The Changing Face of Conjugal Relationships, Catalogue 11-001.

Marriage: Till Death Do Us Part

Marriage is less common in Canada today than it was in the past. However, when you add the increase in common-law relationships, Canadians seem to be pairing up as much as they ever did. The decline in marriage can perhaps be explained by the decline in religious practice in Canada. Some Canadians see their most intimate relationship as a private matter that does not need the blessing of organized religion to be legitimate. Others choose to avoid the legal complications of divorce by not having a legal marriage. However, common-law partners in enduring relationships have most of the financial rights and responsibilities of married partners, especially if they have children. Canadian couples believe that their personal commitment to each other is the most important aspect of their relationship, whether or not they choose to formalize their relationship with marriage.

Divorce

The divorce rate, which increased rapidly through the 1970s and 1980s, is now declining. Changes to the Divorce Act in 1968 enabled couples to divorce for *irreconcilable differences*, that is, they could no longer agree. The increase in divorce can be blamed on our ideal of a conjugal marriage that places an emphasis on love and companionship, not parenthood or economic support. When you no longer enjoy each other's company, or if you do not love each other any more, the marriage is over. Some social scientists have a theory that a higher divorce rate is a necessary side effect of our preference for love marriages between equal partners.

How can you explain the recent decline in the divorce rate? It is possible that those who wait longer to marry are making better choices of a marriage partner. Divorce rates have always been higher for younger people, especially for those who marry as teenagers. Another explanation is that more people are choosing to live in common-law relationships. Common-law relationships are not recorded anywhere when they are formed, nor are they recorded when they are dissolved, so it is difficult to count them. According to recent research on relationships, common-law relationships are less stable than marriages, but the break-up is not visible in the divorce statistics.

Above: Couples must choose whether to make their lifelong commitment to each other in a legal or religious ceremony.

Below: Divorce is more common when people expect their spouses to be their best friend. An end to the friendship usually results in divorce.

CHECK Points ✓

- ✓ What are some possible reasons why fewer people are getting married today?
- ✓ Are people living common-law more or less likely to split up than married couples?
- ✓ What reasons are suggested for the declining divorce rate?

career Link

What is a statistician?

A **statistician** is a person who is trained in the science of statistics.

What do statisticians do?

Statisticians develop and apply mathematical or statistical techniques to solve problems in fields such as physical and biological science, engineering, social science, business, and economics. They use statistical information to produce reports that give important information. Government, business, and industry can use this information. Actuaries calculate probabilities about lifestyle issues and life expectancies for insurance companies. They give information about probabilities so that the insurance company can charge appropriate premiums on insurance policies.

Where do statisticians work?

Business and industry are the primary workplaces for statisticians. They are often key players in a company's future planning initiatives. Statisticians are also employed by governments to calculate probabilities and results of human behaviour. Governments use this information for future planning.

Reflections and Connections

1. Do you think you will marry? What factors will affect your decision?
2. In your opinion, can staying single be as good a lifestyle in Canada as getting married?
3. Why do you think delaying marriage until you are older might improve your chances of staying married?
4. What is your theory to explain that marriages last longer than common-law relationships?
5. What do you think young adults could do to reduce their chances of divorcing?

Working Families

At the end of the twentieth century, the most common form of family in Canada is the dual-income nuclear family. In most families, both parents are employed outside the home. Many people assume that this is a new form of family. However, you have learned that both parents worked in many families before the 1920s. The breadwinner family with one working parent was considered to be the ideal family, but it was only common for a few decades in the middle of the twentieth century. Today, women are just as likely as men to chose an occupation, to plan their education, and to pursue a career.

In the 1970s, two incomes gave the family an economic advantage and enabled them to buy luxury goods and take vacations. By the end of the century, however, the majority of families had two incomes and the advantage was lost. Families became used to buying more consumer goods. Salaries did not increase as much as the standard of living, so a woman's income became as important as the man's in many families for buying the necessities, not the luxuries. Our definition of the necessities of life has changed. The average **standard of living** rose and along with it, the lifestyle expectations of Canadians. As the gap between the income of men and women narrowed, the possibility of economic independence for women in Canada increased. By the 1990s, the Vanier Institute for the Family estimated that it took 70 hours of work per week to support the average family—that is almost two jobs.

Female Participation, Unemployment and Employment Rates, By Marital Status and Age Group, 1999

	%		
	Participation rate	Unemployment rate	Employment rate
Total	58.9	7.3	54.6
15 – 24 yrs	61.7	12.6	53.9
25 – 44 yrs	79.6	6.6	74.3
45+ yrs	39.95	437.7	—
Single	65.4	10.5	58.5
15 – 24 yrs	59.8	13.1	52.0
25 – 44 yrs	82.8	7.7	76.4
45+ yrs	45.7	6.4	42.8
Married	62.7	5.7	59.2
15 – 24 yrs	71.5	10.0	64.3
25 – 44 yrs	78.8	5.9	74.1
45+ yrs	46.1	4.7	43.9
Separated/Divorced	65.5	9.3	59.4
15 – 24 yrs	60.5	—	42.1
25 – 44 yrs	79.8	10.3	71.6
45+ yrs	55.8	8.0	51.4
Widowed	10.2	7.2	9.5
15 – 24 yr	—	—	—
25 – 44 yrs	68.3	12.6	59.7
45+ yrs	9.2	6.5	8.6
- nil or zero			

www.statcan.ca

Young women are as likely as young men to prepare for employment in a career of their choice.

Family income is the major influence on family lifestyle. Families only consume the goods and services that they can afford. An adequate standard of living, which means the cost of the basic necessities, is used as a Low Income Cut Off (LICO) for defining poverty in Canada. In 1998, 15 percent of Canadian families were living below the LICO, sometimes referred to as the poverty line. Although extended unemployment results in poverty, most poor families have working parents. Some forms of family are more likely to be poor than others. Lone-parent families are more likely to be living below the LICO because they only have one adult who

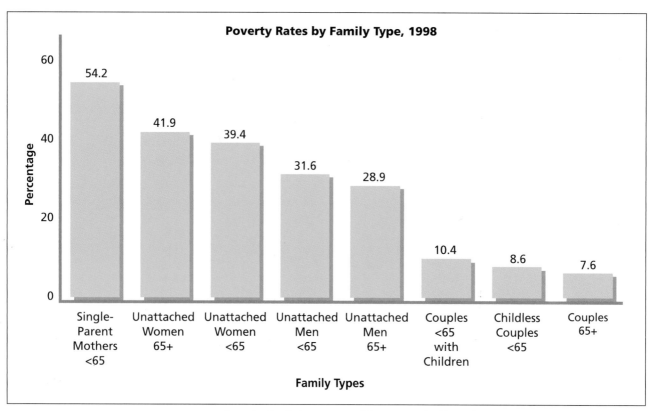

Poverty Rates by Family Type, 1998

Reproduced with the permission of the Minister of Public Works and Government Services Canada, 2001.

CHECK ✓Points ✓

✓ Explain why two-income families no longer have an economic advantage in Canada.

✓ Which forms of family have the highest poverty rates in Canada?

✓ How does Canadian society provide income support for families?

✓ What services are available in your community to help families perform their functions?

is able to earn income to support the family. Because the Canadian lifestyle depends heavily on buying the goods and services that families need rather than producing them ourselves, poverty reduces the ability of families to perform their functions.

Canadian families are able to receive assistance to increase their resources so that they can carry out their responsibilities. Through ***transfer payments*** such as employment insurance and other income security programs, governments provide income for those who are unemployed, unable to work, or earning a low wage. Governments subsidize community services so that families do not pay all the costs. Health care and education are provided to all Canadians as a public service. Community organizations, such as food banks, meals on wheels, recreation programs, and subsidized day care, provide services to families. With help from the community, families are able to extend their income to meet the needs of their members.

Reflections and Connections

1. How would family lifestyles differ if women were no longer employed outside of the home?
2. What do you expect to be your pattern of employment as an adult?
3. What factors will affect whether your spouse will work if you have children?
4. What are some possible factors which result in poverty for families in Canada?

Where Are the Children?

Canadians consider children to be the major part of family life. Many people delay marriage until they are ready to have children and "start a family." Today, effective birth control enables people to plan when to have children. Many Canadians are waiting until they are older, when they feel more settled financially, to have children. Families of two children are most common, but families having only one child are becoming more common. In 1996, 84 percent of children lived in two-parent families, 15.7 percent with a single parent, and only 1 percent lived with someone else. Although there is a perception that "families are broken," most Canadian children are living with their biological parents.

Families With Children in 1996, With Whom Children Live

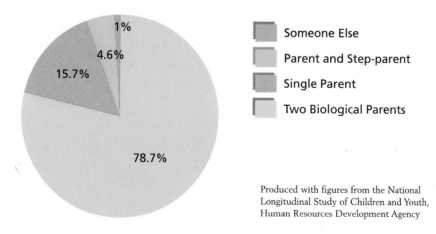

- Someone Else
- Parent and Step-parent
- Single Parent
- Two Biological Parents

1%
4.6%
15.7%
78.7%

Produced with figures from the National Longitudinal Study of Children and Youth, Human Resources Development Agency

When two incomes became a necessity to support our Canadian lifestyle, even mothers of young children needed to go to work. Day care provides care for preschool children while their parents work. In some communities, going to day care is the most common early experience for children. Some children who do not need day care go to nursery school or play groups to acquire the early education that other children get. The extended family might provide the day care in some families, but day care may not be available for children whose parents work shifts. Some parents share the child care, but lose time together as a couple, by working different shifts. Sharing the role of parenting with others outside the family is very common in Canada today.

Some people have difficulty becoming parents. In the past, the only option for *infertile* couples was to adopt a child who had been given up for adoption by its birth mother. In the last thirty years, however, single parenthood, has become more acceptable, even for teenage mothers, so that few babies are available for adoption. Some couples are choosing international adoption, an expensive choice. Reproductive technologies have improved the chances of an infertile couple bearing their own biological child. Some of the newer technology can be very expensive and does not always work. Couples are often willing to go to great lengths to have a child to complete their family.

Since divorce has become more common, the court must decide the custody of the children of divorced couples. **Custody** of children refers to which parent children live with and which parent is the legal guardian. In

Births and Birth Rate*
Birth Rate per 1000 Population

	1994-1995	1995-1996	1996-1997(r)	1997-1998(r)	1998-1999(r)
Canada	13.1	12.6	12.0	11.5	11.2
Newfoundland	10.7	10.4	9.9	9.6	9.4
Prince Edward Island	12.6	13.2	11.8	11.5	11.3
Nova Scotia	11.7	11.6	10.9	10.5	10.3
New Brunswick	11.6	11.1	10.7	10.4	10.2
Quebec	12.3	11.9	11.4	10.6	10.1
Ontario	13.5	13.0	12.2	11.7	11.5
Manitoba	14.6	13.9	13.2	12.8	12.6
Saskatchewan	13.6	13.2	12.8	12.5	12.4
Alberta	14.3	13.9	13.3	12.9	12.9
British Columbia	12.6	12.2	11.6	11.2	11.0
Yukon	15.2	14.6	14.6	14.6	13.9
Northwest Territories	23.4	24.6	22.5	21.5	17.0**
Nunavut	#	#	#	#	27.2**

Note: Data are final unless otherwise indicated.

(r) Updated data

\# Data unavailable, not applicable, or confidential

* From July 1 of one year to June 30 of the next year

** Where available, data are displayed separately for Nunavut to reflect April 1, 1999 partitioning of the Northwest Territories

most cases however, both parents are responsible for the children's support. A hundred years ago, men received custody of children because of their ability to support the child financially. Today, in most cases, custody is granted to the mother because she has been the primary caregiver for the children. As men become more actively involved in parenting, more men are being given custody of children. Canadian law encourages divorcing couples to negotiate shared-custody arrangements in which both parents retain guardianship and the children move between mom's home and dad's home. If mom and dad live close to each other and are on friendly terms, shared custody can be a comfortable arrangement for everyone until the children require more independence.

The two-parent family with two children is the most common family in Canada today.

The National Longitudinal Survey of Children and Youth

The National Longitudinal Survey of Children and Youth is studying a generation of Canadian children as they grow up to determine what their lives are like.

The National Longitudinal Survey of Children and Youth (NLSCY) is part of the federal Child Development Initiative, which was developed in response to the 1990 World Summit for Children. The NLSCY attempts to fill the information gap on children and their development, particularly what happens to children over time. The survey is conducted on behalf of Human Resources Development Canada.

In 1994/95, when the first data were collected, the study included 22 831 children aged from birth to 11 years. The same children will be tracked at two-year intervals into adulthood. This will make it possible both to update the portrait of children in Canada created by the survey, and to gather valuable evidence on the history of young people.

Longitudinal data, information collected from the same people over a period of time, is used to track developmental changes and to study the influence of a child's social and family environment. Information is being gathered using interviews with a parent. If parents give their permission, interviews and questionnaires for older children, interviews with teachers and principals, and testing of vocabulary and mathematical skills are also used in the survey.

The results of each stage of the survey will be released when they have been analyzed and summarized by Statistics Canada.

Statistics Canada, www.stacan.ca

A better balance between work and family is possible

By Rod Beaujot

Rod Beaujot is a professor of sociology at the University of Western Ontario. He has just published Earning and Caring in Canadian Families (Broadview Press).
The Biblical call to Adam and Eve was to "be fruitful and multiply" and to "dominate the Earth." In the usual rendition of this mission statement, it was not for Eve to be fruitful and multiply while Adam was to dominate the Earth. The call was given to the couple, to both men and women. But government policy in Canada often does not encourage a balance of earning and caring between the two genders.

Equal opportunity by gender involves questions of education and work, but also involves everyday life. There has been much change in education, from 1960, when only a quarter of university students were women, to 1996, when women comprised 56% of university students. At work, participation rates have become much similar as well, so that it is difficult to imagine anyone who would not see paid work as a central component of their lives. The transformations at work remain somewhat incomplete, however, as seen in the difference between the things that interrupt a woman's working day—a child's illness—versus that a man encounters.

It is in everyday life that the opportunities of women and men remain particularly uneven. In analysing how this could be different, it is particularly useful to study how men and women use their time. Time provides a common measure for both earning and caring, or paid and unpaid work. Canada has had three surveys of time use and they all show an important result that has not penetrated the national consciousness. When one considers all adults over each day of the year, the average time that men and women spend on productive activities, paid or unpaid, is remarkably similar.

For instance, in the 1998 survey, the averages were the same at 7.8 hours per day for women and men. There are clearly a variety of models for the sharing of earning and caring, with change being both painful and slow. Within couples aged 30-54, 58% are those where one person spends more time at paid work and the other spends more time at unpaid work. One in 20 are men. In 28% of this group, there is a double burden, where both people do the same amount of paid work, but one does more unpaid work. One in four here are men. That leaves 14% of couples where the time use in unpaid work is more equal.

If the earning and caring burden is to become better balanced and women are to have equal opportunities in everyday life, we need to move to models not only of co-providing but also co-parenting.

In Sweden, social policy creates an incentive to postpone child-bearing until careers are established, along with a lesser dependency of women on men and less reliance on joint custody of children. The result is the cost of child rearing is partially transferred from women to men and to the workplace.

Canadian thinking seems to be in the direction of rescuing parents from some of the unpaid work, especially through publicly funded childcare. While that is certainly a possible route, it does not necessarily help redistribute the burden between women and men. Given the considerable interest in work leaves as a way of accommodating the care of infants and young children, how about doubling the current 26 weeks to a full year, with a maximum of half a year per parent?

The London Free Press 03-03-2000

CHECK Points

✓ What are some explanations for Canadians choosing to have fewer children?

✓ What type of family situations are most children living in today?

✓ What are the options available to infertile couples who want to have children?

✓ According to Canadian law what is the preferred custody arrangement for children after divorce?

Reflections and Connections

1. How do you think that family lifestyles would be different if the trend towards one-child families grows?
2. What do you think children in day care learn that other children do not learn?
3. How do you think families would benefit from a national affordable day care program?
4. Why are so many teenage mothers keeping their babies rather than giving them up for adoption?
5. Should reproductive technologies be available to all women or only to married women?

Health Care and the Family

Improvements in health care have helped to strengthen the Canadian family. Improvements in prenatal and obstetrical care have greatly reduced the infant mortality rate in Canada so that most babies born today will live to become healthy adults. Immunization of children against infectious diseases has almost eliminated epidemics, such as diphtheria and whooping cough, which used to kill many children. Antibiotics also improve the chances of recovering from infections. Knowing that children will survive into adulthood changes the relationship between parents and children. Over the last century, Canadian parents had fewer children and developed a more loving relationship with each child as an individual. Today, parents can be confident that their children will probably live a long and healthy life.

The health of adults has also improved. The *life expectancy* for adults in Canada has improved from 59 years in 1920 to 78 years in 1990. A longer life span has changed the lifestyle of Canadians. As you know, Canadians are marrying later, and having their children later. Young adults are in no hurry to choose a job, a home, and a life partner and settle down. There is no rush. Adults can look forward to many years of life after their children leave home and years of leisure after they retire from their jobs. Having a longer life requires that individuals plan their finances carefully when they are young so that their pension will enable them to enjoy a comfortable

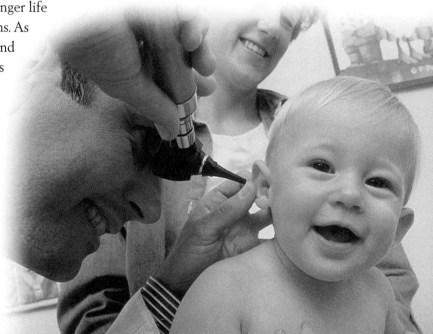

Today, Canadian children can look forward to a long and healthy life.

Aimed at women who want children later in life

By Idella Sturino

Montreal—A researcher at McGill University hopes to develop a new birth-control pill that would also delay menopause—improving women's chances of getting pregnant up to age 50.

Dr. Roger Gosden, director of McGill's reproductive biology division, wants to design a contraceptive device—being dubbed the "career" pill that would pause the aging process of women's ovaries.

The research is at a preliminary stage and it could be 15 years before it is completed, but some people are already raising questions about the project.

Gosden said yesterday that many women are putting off having children until their mid-30s to late 40s while they develop their personal relationships or careers. But as women age it becomes harder for them to conceive and many wish they could delay menopause.

"We wonder whether there is a way to address all those issues," Gosden said at an impromptu news conference outside the Royal Victoria Hospital in downtown Montreal.

"There are all sorts of social pressures for people to extend their training periods (at work) and people are settling down with partners later. Yet the biological limits of our fertility are fixed."

Current birth-control pills tell the ovaries to stop eggs from developing fully and they are discarded.

The contraceptive method Gosden wants to make—which could take the form of a pill, spray or injection—would delay ovulation, storing the eggs for use later once the woman stopped using it.

Gosden stressed it remains unclear whether the eggs stored for late use would lead to healthy babies.

"The question remains of whether those eggs have the same quality of those of younger women, or are more liable to produce an embryo which doesn't thrive and results in a miscarriage."

Gosden, who would not reveal where the funding for his research is coming from, said he's not out to encourage women over age 50 to have kids, only to offer an alternative to women who might have fertility problems in their 30s and 40s.

But not everyone is impressed with the idea.

Abby Lippman, a professor of epidemiology and biostatistics at McGill, said the answer to women wanting to balance a career with child rearing doesn't lie in science.

"Women don't want to delay their childbirth," she said yesterday after being told about Gosden's research.

"They want to be able to have their working lives respected and their mothering lives respected when they have the energy and the time. It's up to society to allow women to have both, not doctors."

The Toronto Star, July 21, 2000
Reprinted with permission of The Canadian Press

What do you think?

1. Why might it be difficult for a woman to have children and a career in her 30s or 40s?

2. What are some advantages and disadvantages of delaying child rearing until her 50s?

3. What does Dr. Lippman mean by, "It's up to society to allow women to have both, not doctors?"

4. Some people think birth-control pills are immoral. Do you think this method would be more acceptable because it would work differently?

lifestyle in old age. With careful planning, a long life expectancy allows us to experience different stages in our lives and enjoy different lifestyles.

Perhaps the most significant change in health care in Canada has been the introduction of health care programs funded jointly by the federal and provincial governments through taxation. Universal health care ensures that all Canadians have access to the medical treatment that they need regardless of income. Families do not pay directly for their medical care, but pay indirectly through taxation. The financial burden of medical care for serious illness has been lifted from individual families. There is growing concern that the cost of providing medical care for Canadians is becoming too expensive, but deciding how to reduce the cost means deciding how medical care will be reduced.

CHECK Points

✓ Why does a lower infant mortality rate result in smaller families?

✓ How would a longer life expectancy change family life?

✓ How has Canada's health care system improved the quality of life for Canadians?

Business and Families

Families and businesses have a reciprocal relationship because they need each other. Businesses need families to buy the goods and services that they produce. They also need to employ family members in order to stay in business. Families need the income they earn by working at a business in order to buy what they need. In the past, as you have learned, families were expected to produce everything they needed. Now, most people work at a job to produce one product or service, and purchase almost everything else. The demand for new products, such as cell phones and pagers, results in new businesses that make, advertise, sell, and service the products, thereby creating new jobs. Families purchase more, and jobs are created. These jobs enable families to spend more—a cycle of increased spending and increased income.

A longer life span enables Canadians to enjoy an active lifestyle with their grandchildren in their retirement.

Financial institutions are businesses that provide financial services for families. These services can often be used to improve the quality of family life. Families can also invest their money with financial institutions to earn interest. Investment can increase the family income. Banks, trust companies, and credit unions lend money to families in the form of mortgages that allow them to buy homes. They also extend credit so that people can use their credit cards to buy and use products today, before they can afford to pay for them, and pay for them later. There is an invisible added cost because retailers add the credit costs to their prices. Also, interest on credit card perchases is very high.

Changes in the way that business is conducted have had an impact on family life. For example, in many towns and cities in Canada, stores are open until 9:00 P.M. each day and on Sundays as well. These hours allow flexibility for family shopping, but also require that retail employees work late into the evening and on weekends instead of being at home with their families. On the other hand, some businesses have introduced flextime, programs that allow employees to decide their own hours each day within the limits defined by the company. Flextime accommodates the needs of parents who need to be with their children.

Above: Financial services such as mortgages allow many families to borrow money to purchase their own home.

Right: Businesses provide the items that families want to buy and provide employment for family members.

The separation of work and family life appears to be reversing. More Canadians are working at home now that computers enable us to conduct our business over the Internet. This is beneficial for companies because it reduces the need for office space. Self-employed people find that reduced commuting time extends the time available for productive work. The home office offers flexibility in working hours that can benefit families with young children. Also, children who contribute to the work by photocopying, stapling, and stuffing envelopes for mom or dad are learning about how business is done. Working at home requires strict self-discipline to start work, to avoid distractions during the workday, and to stop work at the end of the day. Trace this trend, as it develops, to see the effect on family life.

Shift Work Alters Family Life

By Elaine Carey

For most of the week, Paul and Donna Wyatt are like two ships passing in the night.

Donna tiptoes out of their Oshawa home at dawn to work the 6 A.M. to 3:30 P.M. shift at Hallmark Cards in Toronto, then leaves to pick up baby Jessica just as Paul arrives to start his 3:30 to 11:30 P.M. shift. He climbs into bed around 2 A.M.

"I miss my wife and she misses me, but nowadays you have to do what you have to do,'" says Paul, 27, a stock handler at Hallmark. "We both know babysitting fees are just unbelievable.'"

For more and more families, shifts have become the norm, Statistics Canada said yesterday in its first look at the schedules of working couples. At least one partner had shift work in 634 000 couples, or four out of 10 of the nearly 1.7 million working couples in Canada in November, 1995, StatsCan says. One in four husbands and one in five wives do shift work, and about 40 000 couples are working at completely different times of the day.

What's disappearing from the schedule is time together. While couples who work an average 9 A.M. to 5 P.M. day have about 16 hours left to be together, shift couples have four hours and 20 minutes less because their jobs overlap and their workdays are longer. While most say shift work is part of the job, one in 10 say they're doing it for other reasons—for husbands, it's largely to make more money, for wives, it's to take care of the children.

All the blue-collar occupations report above-average rates of shift work, as do jobs in medicine and the service industry, the report says. Men outnumber women except in the fields of nursing and sales.

Younger husbands and wives and those with preschoolers are the most likely to be working shifts, supporting the theory they're staggering their schedules to care for the kids, the report says. For the Wyatts, "it's so much easier, it's the best way to do it," says Paul, who cares for 19-month-old Jessica until he drops her at a babysitter at 3 P.M. His wife picks her up an hour later. "If I can spend quality time with my daughter, and my wife can, everybody's happy," he says.

Shift work is the outcome of globalization and "the perceived need to be able to do business 24 hours a day," says Bob Glossop of Ottawa's Vanier Institute of the Family. Add to that the growth of the service sector, with stores open in the evening and on Sundays and "you have a different kind of labour force today," he says. But what's good for business is not necessarily good for the home, he warns. "What it means for families is they've become ships passing in the night. In my opinion, time is the essential building block of young families."

But while parents may choose to work different shifts in order to spend time with the kids, what's in jeopardy is quality time between the parents, Glossop says. "They're passing each other notes or making phone calls at the end of the shift saying, 'Don't forget to take the pork chops out of the freezer and take the kids to the dentist in the morning,'" he says. "They can't even catch up on weekends because they often have to work them as well." "From a family point of view, it is a severe challenge," he says. "It's a practical response to make sure somebody's at home to provide childcare, but the relationship is in jeopardy. We have to dedicate time to each other as well as our kids." But most young families don't have any choice, he adds. "They're going to the mall and working from 4 to 9 at night at Zellers, trying to get enough out of the job to put a meal on the table."

The Toronto Star, October 9, 1998
Reprinted with permission—
The Toronto Star Syndicate

It can be difficult to separate work and family if you work at home.

✓ What new businesses have emerged to provide goods or services for families?

✓ How do mortgages help families to buy a home?

✓ How can the use of credit improve a family's standard of living?

✓ Explain the challenge of working productively at home.

Reflections and Connections

1. Do you think families would be better off if they just cut back on their buying?

2. What goods or services do the working members of your family produce?

3. How have the changes in business and working hours affected your family?

4. If more and more Canadians work at home, what do you predict will happen to family lifestyles?

Summary

1. Some reasons why Canadians are delaying marriage are that they stay in school longer, both men and women expect to work, and many accept sexual relationships outside of marriage.

2. Most Canadians form committed life-partnerships in marriage and, increasingly, in common-law relationships.

3. In Canada, the dual-income nuclear family has become the most common family for maintaining a comfortable standard of living.

4. When both parents are employed outside the home, day care provides substitute care for preschool children.

5. Social policies provide for financial assistance and subsidized services to support Canadian families.

6. Most children in Canada live with both biological parents, or in shared custody arrangements.

7. Universal health care and medical advances ensure that most Canadians will live a long and healthy life.

8. Businesses and families are interdependent because increased employment income leads to increased spending.

In this chapter, you learned to

- compare the roles of family members in a variety of family forms;
- distinguish between facts and opinion about the current state of the Canadian family;
- summarize the impact of social, economic, legal, technological, and environmental changes on the lifestyles of individuals and families;
- analyse the impact of social and political policies on the lifestyles of individuals and families; and
- conduct research about family issues using charts, graphs, and surveys.

Activities to Demonstrate Your Learning

1. Use E-stat to locate the Statistics Canada data on the forms of families in your school community from the most recent census. Compare the data for your community with the data for all of Canada, and suggest reasons for any differences.

2. Using social science research methods, design and conduct a survey to determine the attitudes towards marriage of senior students in your school community and adults. Summarize the results and write a brief report.

3. Tabulate the types of relationships portrayed on television in prime time. Analyse whether the portrayals reflect the reality of Canadian families.

4. Prepare for and conduct a class debate on the resolution, "Be it resolved that marriage is an outdated institution that is no longer necessary."

Enrichment

1. Investigate the types of jobs that can be performed from a home office. If possible, use the Internet as your primary resource. Research the education, skills, and preparation required for these jobs. Compare income possibilities of working from home and make sure to include cost saving benefits such as reduced transportation costs. Present you findings in the form of a career path report.

2. Do some statistical analysis. Find out the rate at which life expectancy is increasing and project approximately how long you can expect to live. Using the same measure, estimate when you might have children and then project what their life expectancy might be. Consider some of the implications to families as life expectancy continues to increase. Present your analysis, projections, and future considerations.

Contributing to Your Family

Where do you fit into your family? How has your role changed since you became a teenager? Are you beginning to move away from your family? Are you choosing to spend more time at school and with your friends? Negotiating your new position in the family can be your first step towards achieving independence. But how do you start?

By the end of this chapter, you will be able to

☀ describe strategies for forming effective family relationships;

☀ describe the various ways in which family members contribute to the family;

☀ explain the various ways in which teenagers are responsible within families; and

☀ summarize strategies for negotiating your responsibilities and privileges within a family.

You will also use these important these terms:

authority	responsibility
negotiate	sibling rivalry
power	volunteering

Changing Family Relationships

When you were an infant, your family relationships were decided for you. You did not have a say in who your parents were, or whether or not you had siblings. Nor could you control whether the relationships in your family were happy ones. Your family probably provided you with love and support during your early years. Now that you are an adolescent, you expect more from your relationships, even within your family. You can choose to develop your family relationships to meet your needs on a reciprocal give-and-take basis. In other words, you can give to your family as well as receive benefits from it. What can you offer to your family as you take on more responsibility for your own life?

A hundred years ago, your role would have been clearly defined. As a teenaged boy or girl, you would have known exactly what you would be doing at home, at school, and in your future. Today, in our diverse society, the roles of males and females, young and old, have changed. Families differ in their expectations, and each of you must *negotiate* your position in your family, in your peer group, and later, in society as an adult. One thing is clear—when you reach adulthood, you will be expected to be independent and responsible. You are already beginning to develop independence and responsibility as a teenager.

In the past, your role was clearly defined. A young man knew exactly how he would be expected to live as an adult.

It is unjust to claim the privileges of age and retain the playthings of childhood.

Samuel Johnson

All happy families resemble one another; every unhappy family is unhappy in its own way.

Leo Tolstoy

Strong Families

You have learned that effective families are those families that perform their functions well. Many sociologists have examined the family to find out more about what makes a family effective. They wanted to find out why some families are successful at meeting the needs of their members even when they face difficulties. Their research concluded that effective families:

- are able to manage their resources to provide for the well-being of all family members;
- delegate *responsibility* for the work necessary to maintain the family;
- have a clear power structure so that everyone knows where he or she stands; and
- are fairly efficient at accomplishing the work of families.

All families have problems to solve. Families face challenges as they move from one stage to another. For example, all couples have to negotiate how they are going to resolve conflict and make decisions. Besides common daily problems, more difficult circumstances may produce added challenges for families.

Serious illness or disability, unemployment, political strife, war, and migration, are strenuous circumstances some families have had to deal with. Some families are resilient and survive these difficulties, but others do not. By the end of the twentieth century, about one in three Canadian marriages ended in divorce. This breakdown of families results in additional challenges for parents and children as they attempt to meet their individual developmental needs. Effective families are those that manage the crises that occur in family life in as positive a way as possible.

The members of strong families support each other emotionally and socially. This means that family members respect each other as individuals and accept their differences. You have learned that clear and open communication helps individuals to express their thoughts and feelings, to understand each other and to feel empathy. Another aspect of respect in families is for parents to allow their children to make mistakes and learn from the consequences. Respect allows individuals to develop responsibility for their own lives. Strong families express genuine warmth and affection for each other.

Family Time

Strong families spend time together. As in all relationships, close family relationships develop when people make the commitment to be together in order to get to know each other better, and to share experiences. Family mealtime allows family members to talk about their lives and share their ideas informally on a daily basis. Research has shown that children who eat meals with their families are more emotionally stable and are more successful at school. They often develop habitual ways of communicating and sharing activities that become family rituals.

Family togetherness provides a stable environment for children and teenagers in all stages of the family life cycle. As a teenager, you can expect to leave your parents' home eventually in order to become independent and, for

Shared custody is one way of enabling divorced parents to provide an effective family for their children.

Let us act on what we have, since we have not what we wish.

Cardinal Newman

Zits

most of you, to form your own family. Before your parents become an "empty nest family," you must "develop your wings and learn to fly." Unlike birds, however, you know that you will be leaving. To make it less painful, people have learned to leave home gradually, by loosening their ties to their families over time, and by developing relationships outside the family to replace the support of the family. It should not be surprising to note that, statistically, teenagers do not rate their family life as important as do children or adults.

The Support You Need

Not all children and teenagers live in strong families. In some families, the adults have their own problems to cope with. If they are unable to manage these problems, they may have trouble maintaining their relationships with spouses and even their children. If the problems are temporary, the relationships might improve. Unfortunately, some troubles are long lasting and the family does not recover. Because a strong family environment is so important for individual development, children and teenagers from troubled families often need support from other people or organizations. Having an understanding of your own needs can enable you to get the support you need. Understanding what others in your family need could mean that you provide support beyond what would normally be expected from someone your age.

CHECK Points

- What are the characteristics of strong families?
- What are the benefits of spending time together as a family?
- Explain why teenagers need to spend social time away from their families.

Reflections and Connections

1. Discuss the ways in which you are beginning to loosen the ties to your family in order to become independent.
2. In what ways might crises affect how well families are able to perform their functions?
3. How do strong families manage crises?
4. Describe situations that you are aware of where teenagers provide extraordinary support to their families.

Family Conflict

Strong families have a clear power structure. **Power** is the ability to control the behaviour of another. You have power when someone is willing to do what you want him or her to do. Power is a good thing because power enables you to get what you need in life. However, the misuse of power results in unhealthy relationships. A power balance exists in relationships where there is give and take. In a reciprocal relationship, each person has the ability to meet the other person's needs in some way. In equal relationships, both people have equal benefits to offer each other and so the power is shared evenly. In unequal relationships, one person has more to offer than the other does. The person who has more to offer has more power; the person who needs more has less power. In families, it is clear that some members have more to offer than others do and thus have more power.

Power in Relationships

When an individual has the ability to meet another's need or to withhold something the other needs, the individual has power.

Financial Power	The ability to provide or not provide money or the things that money can buy.
Physical Power	The ability to protect or, potentially, to inflict physical harm.
Emotional Power	The ability to provide love or affection, or to withhold love.
Expert Power	The ability to use knowledge or skills to solve problems or to create problems for others.

power quiz

Who has the power in each of these relationships?

Veronica really wants to have a boyfriend. She has been going out with Jason for a few weeks now, and she really likes him, but she heard that he was out with Kim last night. Jason wants to come over tonight when Veronica's parents are out, but Veronica knows her parents will be really angry if they find out.

Sameer has been assigned to work on a project with Franco. Sameer does not like Franco although he is not sure why. Franco will be a good partner on this project because he is the best student in the class, and Sameer is struggling to keep up. Franco looks at the list of topics and suggests a difficult topic. Sameer wants them to choose another topic.

Nicole would like to go shopping at the mall on Saturday afternoon with her friend, Sandy, but she needs a ride. She asks her father if he will drive her. He answers that he will take her when she has finished vacuuming the house as she said she would do on Saturday. Nicole replies that she was hoping to sleep in. Now, she would have to get up earlier to vacuum.

The arcade is crowded after school. Ivan is playing at his favourite game. He is doing really well and might break his own record. Steve comes up behind him and starts jostling him and demanding a turn. Ivan looks up at Steve, who is much bigger, and asks him to stop.

Conflict between teenagers and parents is inevitable. What differs from family to family is how the conflicts are resolved. The major task of adolescence is to develop an individual identity. As you strive to discover who you are, you will learn that although you have many similarities, you are not your parents. In fact, you can be very different from your parents. Parents might interpret this understanding as a rejection of who they are as individuals. This rejection is often the root source of conflicts between parents and teens. As you become aware of your strengths and weaknesses, your interests and talents, and look for opportunities to develop these, you will spend more of your time away from your family. You become less dependent on your parents to support your self-concept, and rely more on your own judgment.

You are also asserting your independence as you develop your own "wings." You will probably make some decisions differently from the way that your parents would make them. Whether you are choosing how to wear your hair, what courses to study at school, what friends you will hang out with, or your future career, your decisions reflect the person you are becoming as you mature. The maturing you is not always the person your parents would like you to be. Remember that conflict can develop out of power struggles, personality differences, and emotional situations. You can see that parent–teenager relationships have many possibilities for conflict.

Resolving Conflict

Conflict with your parents can be more difficult to resolve than conflict with your peers. Your relationship with your parents is probably not an equal relationship. Parents have the legal *authority* and responsibility to ensure that you are well behaved and make wise choices until you reach the age of majority at 18 years of age. They are also obligated to support you until at least 16 years of age or until 18 years of age as long as you "remain under their authority" in legal terms. In most cases, your parents support you financially, as well as socially and emotionally. In terms of the social exchange theory, the benefits of your relationship with your parents outweigh the costs. You need them more than they need you. In this unequal relationship, you may find that you resolve conflict most often by giving in or compromising.

How parents wield their power reflects their different parenting styles. Some parents allow you to make your own decisions from an early age and may set very few limits on your behaviour. These permissive parents can vary from easy-going "buddies" to negligent parents who fail to carry out their responsibilities. At the other extreme are parents who make all of the decisions about your life and attempt to control every aspect of your behaviour. These authoritarian parents might be rigid disciplinarians, or may simply be trying to protect you from disappointment and harm. Democratic parents, who share the decision-making with you whenever appropriate, and who allow you to suffer the consequences of poor choices, occupy the middle

How to Get Along with Your Parents

Spend time together sharing activities that you all enjoy.	Sharing your time is part of sharing yourself with others and letting your parents share their lives with you.
Communicate with your parents.	Communication goes both ways, so talk to them about your life and listen to them talk about theirs.
Treat them with respect.	Try to understand their thoughts, feelings, and needs and take them seriously.
Express your feelings.	Let them know that you appreciate the things that they do. If you love your parents, find ways to tell them.
Negotiate solutions.	When there is a problem, look for solutions that meet your needs and their needs. Explain your concerns and listen to their advice when the problem is yours to solve.
Be responsible.	Do what you say you will do, clean up your own messes and apologize when you make a mistake that affects others.
Be nice!	

ground. All styles can work. Very few parents want to make their teenagers miserable. Identifying the parenting style of your parents can help you understand their position.

You will be able to negotiate with your parents as you mature and your relationship becomes more equal. You gain power by becoming more responsible in your family. Many teenagers discover that as they become more responsible within the family, they are able to gain more independence. As they make some decisions successfully on their own, their parents begin to trust teenagers' judgment. Your success at negotiating with your parents to resolve conflict or to get something you want will depend on your parents' view of you. They need to have trust in your ability to make the right choice and to accept the consequences yourself. Taking responsibility can lead to more independence.

Zits

CASE STUDY

Claire's Story

When Claire arrived home from school after choir practice, she was surprised that her younger brother, Kyle, was not home. Kyle had not done his job of cleaning up the breakfast dishes, either.

Claire's mom would be home in a few minutes to make dinner and would be annoyed. She had told

Kyle the other evening that he had to make sure he did his job because she depended on him to have the kitchen clean so she could prepare dinner.

Claire checked to see if there was a message from Kyle. There was not.

What do you think?

1. What would be the best action for Claire to take:
   a) start her homework because dinner will probably be late?
   b) clean up the breakfast dishes so that mom can start dinner?
   c) wonder where Kyle is as she grabs something to drink and heads off to watch television?

CHECK Points

✓ Describe the sources of power in family relationships.
✓ Explain the possible causes of conflict in parent-teenager relationships.
✓ How can a teenager gain more power in the family?

Reflections and Connections

1. How has your role in the family changed since you became an adolescent?
2. Think of a conflict situation that recently occurred in your family. What was the power balance in the situation? How was the conflict resolved? How could it have been resolved differently?
3. What benefits do you receive from your family? What do you have to offer that would be of benefit to your parents or the rest of the family?

Your Place in the Family

Each member of the family has a different role to play. Although siblings live in the same family, brothers and sisters do not have the same siblings, so they all, in fact, live in slightly different families. Birth order theory suggests that your position within your family determines the challenges you will face as you grow up and the expectations that others will have of you. This theory suggests that your personal motivation to achieve, your skills in negotiating what you want, and your level of obedience to authority might depend more on birth order than any other factor. It is important to remember, however, that each family is unique. Other factors such as a wide age gap, multiple births, divorce, or remarriage can change the impact of birth order. Your position in your family will affect what your parents expect of you and the responsibilities you will have in your household.

Your role in the family is also affected by your gender. Some cultures place distinct role expectations on girls and on boys. In Canada, there are very few legal distinctions, and even those are being challenged in the courts. Boys and girls may have the same rights, but most families still expect boys and girls to be different. Families that believe that girls are not as strong as boys might restrict the sports girls can play, whether or not they do outdoor chores, or even their career choices. Family expectations can sometimes be in conflict with those of the public school system. Schools encourage both boys and girls to participate in a variety of activities and require boys and girls to take the same mandatory subjects.

Generally, families tend to be more protective of girls, and to have higher expectations of boys. Historically, in many cultures the underlying reason for differences in dress code or curfews for girls was the desire to protect girls' virginity. Traditionally, boys were expected to be ambitious so that they

Birth Order

Psychologist Alfred Adler explained that each child in a family has a different psychological situation that determines the child's personality.

Birth Order	Family Situation	Personality
Only child	Parents have no experience being parents. Receives full attention of parents, but is alone a lot	Likes being the centre of attention, but might become dependent upon this attention. Has difficulty relating to peers. Likes to be independent
First Born	Parents have high expectations. Often given responsibility for younger siblings	Tries to do well to please because they are the Guinea pigs. Might become bossy. Learns to share everything. Likes to be an expert
Middle Child	Always has someone ahead, setting the pace. Is caught between oldest and youngest and may feel left out. Parents are more relaxed.	Compares self with others. Might feel unable to do well enough to get attention. Might have to fight to find a place. Seeks justice
Youngest	Has many people looking out for him. Pampered by siblings	Often overindulged. Might seem immature forever. Has big ideas but has trouble getting things done

What do you think?

1. Would you describe your family situation in the same way as Alfred Adler?
2. Do the patterns suggested by his theory match your personality?
3. Do the patterns described by Dr. Adler match the personalities of your friends?
4. What are some other birth orders? How would the family's psychological situation affect personality for them?

could support a family. Families of all boys or all girls tend to have fewer limits based on gender, but growing up with a sibling of the opposite sex can give you some insight on "how the other half lives." Because gender roles are based in deeply held cultural values and beliefs, deciding what role you want to play can be very challenging for some teens. Deciding whether you will accept or rebel against the role expectations your parents place on you is part of forming your identity.

Sibling Rivalry

"Someone has got to help me. I think my parents hate me. My little brother never gets in trouble for anything even though he starts most of the fights. I am going insane with him in the house and my parents are not helping. They do not talk to me about anything. They just tell me I am being immature. My whole life was perfect until he was born. Please help!"

When "Angel Wings" posted this message on an Internet chat room, she was experiencing *sibling rivalry*, a normal part of family life. Children naturally compete for their parents' love and attention. The rivalry arises out of the different roles children acquire within the family. Even well meaning parents who try not to compare their children but, instead, support their uniqueness, describe their children as "responsible," "a great skater," "smart," "our little beauty," or "Mommy's big man." Parents might think they are recognizing your strengths, but kids know better! You might feel that if he is "smart," you are not. For many brothers and sisters, parents may have set up the competition to win all the attention by being better at something.

What your parents expect of you will be affected by your gender and your position in the family.

Although the world is full of suffering, it is also full of the overcoming of it.

Helen Keller

You cannot be better at everything, and that is the problem that causes conflict between many siblings. Some people, young and old, have difficulty recognizing someone else's accomplishments. There are several approaches to the sibling-rivalry problem. One way is to feel jealous, and either put down the sibling or dismiss what they do as unimportant. Another way is to feel discouraged or to compete. Some people who feel discouraged try to bring the sibling down in their parents' eyes by getting them in trouble. It is not easy getting beyond your own feelings to see that your sibling also feels discouraged or jealous.

A constructive solution to sibling rivalry is to recognize that the competition does not have to continue. Having positive self-esteem frees you from engaging in the competition. The key to developing self-esteem is to know yourself, your strengths and weaknesses, and to accept who you are. The key to healthy relationships is to accept others for who they are. Once you understand that you and your sibling are probably different in many ways, and do not need to compete, you are on the road to improving the relationship. "Angel Wings" cannot go back to being an only child and the centre of her parents' world, but she can learn to be an older sister and to try to see what her little brother wants in his relationship with her. Once she stops competing for attention, she might gain some independence and some peace.

Reflections and Connections

1. Describe your position in your family.
2. What are the expectations concerning your behaviour that are based on your gender? How has your family expressed these?
3. If you have a sibling, identify any events that have indicated sibling rivalry.

Pitching In to Help Out

Maintaining a household and caring for family members is a lot of work. Recent studies by the Vanier Institute for the Family suggest that it takes about 70 hours of paid work to earn enough money to support the average Canadian family lifestyle. But how much unpaid work is required to keep the household running smoothly? How long does it take to shop for the family's needs, prepare meals, wash dishes, clean the toilet, shovel the snow, vacuum the carpets, dust, cut the grass, change the light bulbs, do the laundry, etc? No one knows the answer, perhaps because no one ever finishes the job.

Families differ in their expectations about housework. What household tasks should be done? How should they be done? How often should they be done? What are the standards for clean rooms, tidy closets, appetizing meals, or attractive clothing? Families have to ensure that their members are fed, clothed, and have a bathroom and a place to sleep—these are basic needs. All of the other things are optional lifestyle choices. No one ever died, or even got sick, from an untidy home. Housework reflects the lifestyle choices and the values of the members of the household. Choosing not to meet the standards set by your parents means challenging their lifestyle and their values.

career Link

What is a home economist?

A home economist is a person who deals with the science and art of managing daily living.

What do home economists do?
Home economists provide knowledge and skills to assist individuals and families in improving their quality of living. They inform their public and corporate clients about the possible consequences of services, products, and policies, which may impact the well being of individuals and families. Home economists work in all areas that assist families in meeting their basic needs. The areas they work in include: human development and family relations, lifestyle and resource management, food and nutrition, clothing and textiles, and housing. Home economists can work as childcare counsellors, district home economists, community nutritionists, fashion consultants, and ergonomics advisors.

Where do home economists work?
Home economists can work in government agencies as managers or policy analysts. Within community agencies, they may work with the elderly or families. In business and industry, home economists can work as fashion merchandisers, public relations specialists, consumer relations officers, or product researchers. There are many freelance opportunities for home economists in all areas of the public and private sector. Many home economists work in the field of education as family studies teachers.

Housework

A hundred years ago, housework was clearly defined. Families produced more of their own goods and services back then. Bread had to be baked, clothes washed and hung to dry, chickens plucked and cooked. Each day had its scheduled tasks to be accomplished. Monday was washing day—it took all day. The technology available today has made many of the jobs easier, and we can buy most of the things we need instead of making them. We work to earn the money to buy the goods and services because we believe that they are necessary for an acceptable standard of living. Does the advertising to influence our buying habits create higher standards than are necessary?

Do you know anyone who enjoys doing housework? Housework is less valued because it is unpaid work, but it is valuable to the family. Living in a well maintained home, eating meals that are appetizing, and being able to express yourself through your living space contribute to the quality of life for you and your family. How much time and energy can you contribute to making your household a practical, clean, and comfortable environment?

Pitching In Pays Off!

Make a meal.
Cooking can be an opportunity for you to develop your skills. Planning what to prepare and shopping for the ingredients help you learn time and money management skills. Cooking a meal gives you some control over what you eat and the quality of your food. As a bonus, everyone who loves to eat loves a good cook.

Look after the children.
Spending time with the younger children in your family will enable parents to do something else as you develop your communication and organizational skills and your perseverance. You also have a chance to play and do childish things that you can no longer do unaccompanied by a child.

Do the laundry.
Doing your own laundry helps you make better decisions when shopping for clothes. You can ensure that your clothes are washed, dried, and ironed to your standards. You will learn which fabrics and features to avoid because they require too much work to maintain. Toss in someone else's clothes to make a full load and you will brighten their day.

Tidy and clean.
Cleaning and tidying can make your home more comfortable to live in, especially if there are members of your household with allergies. Cleanliness can brighten everyone's mood. You will be able to find things more easily, and save time for other things. The bonus is that cleaning and vacuuming can be good way of releasing stress and even provide some exercise.

Average time spent on activities,[1]
total population and participants, by sex

1998		
	Total Population [2]	Participants [3]	Participation rate [4]
	Both sexes		
	Hours per day		Percentage
Paid work and related activities/unpaid work	7.8	8.0	98
Paid work and related activities	3.6	8.3	44
Paid work	3.3	7.7	43
Activities related to paid work	—	0.6	8
Travel	0.3	0.8	38
Unpaid work	**3.6**	**3.9**	**91**
Household work and related activities	3.2	3.6	90
Cooking/washing up	0.8	1.0	74
House cleaning and laundry	0.7	1.7	41
Maintenance and repair	0.2	2.5	6
Other household work	0.4	1.3	30
Shopping for goods and services	0.8	1.9	43
Primary child care	0.4	2.2	20
Civic and volunteer activities	0.4	1.9	18
Education and related activities	**0.6**	**6.2**	**9**
Sleep, meals, and other personal activities	**10.4**	**10.4**	**100**
Night sleep	8.1	8.1	100
Meals (excluding restaurant meals)	1.1	1.2	92
Other personal activities	1.3	1.3	95
Free time	**5.8**	**5.9**	**97**
Socializing	1.9	2.9	66
Restaurant meals	0.3	1.6	19
Socializing (in homes)	1.3	2.4	55
Other socializing	0.3	2.6	12
Television, reading, and other passive leisure	2.7	3.2	85
Watching television	2.2	2.8	77
Reading books, magazines, newspapers	0.4	1.3	32
Other passive leisure	0.1	1.1	9
Sports, movies, and other entertainment events	0.2	2.7	6
Active leisure	1.0	2.4	40
Active sports	0.5	2.0	24
Other active leisure	0.5	2.3	22

— Nil or zero

1. Averaged over a seven-day week

2. The average number of hours per day spent on the activity for the entire population aged 15 years and over (whether or not the person reported the activity)

3. The average number of hours per day spent on the activity for the population that reported the activity

4. The proportion of the population that reported spending some time on the activity

Statistics Canada, *General Social Survey, Cycle 12: Time Use* 1998.

Part-Time Jobs

Getting a job enables you to contribute to your family as well as develop some skills that you will need in the future. Earning your own money reduces the amount of money that you expect your family to spend to support you. Nowadays, most teenagers who work keep the money they earn and control how it is used. Buying your own things or saving to pay for things later with your own money is the best way of learning money management skills.

Whether you get a part-time job is a decision that should be made carefully. As with most decisions, there are various alternatives to consider. Opportunities to earn money will vary depending on the community you live in. You must also consider the consequences of the alternatives not only on your income, but also the impact on your schoolwork, your social life, and your other activities. If you spend some of your time earning money, you will have less time for the other parts of your life. Because you will also have less time for the unpaid work you are contributing to your family, your job will affect other family members. Make a good decision about work by considering the big picture.

The benefits of working are more than just financial gain. Work experience helps you with the task of preparing for a career. Working enables you to develop the specific skills needed on the job, such as following instructions, communicating with employers and customers, and persevering to get the job done within the standards expected. Working at a variety of jobs lets you test your interests, as well as your strengths and weaknesses.

Although working part-time can have many benefits, it has some potential risks as well. Safety in the workplace is a concern for all working people and should be your concern when you work part time. All work carries some risk; even serving ice cream can cause wrist injuries. You have the right to work in a safe environment. You have the right to know about any possible unsafe materials or equipment used in your job. You have the right to participate in health and safety training to avoid injury. If you suspect that you have been asked to do something that is unsafe, you have the right to refuse without being fired or docked pay. Provincial and territorial laws protect these rights.

CASE STUDY

Adam's Story

Hi! I'm Adam. I have enjoyed Grade 9. I found the work in my courses easier than I expected, except in science, but even my science mark has improved this term. I've got to work harder in science next year. I have to study science at university in order to go to medical school to become a doctor. I've wanted to be a doctor since I broke my wrist in Grade 6. I was in hospital for a few days because I had an operation. I was fascinated by what the doctors did and I knew right then that I wanted to be a doctor. But I know I have to work hard to get there.

High school has provided lots of other opportunities for me. I tried out for both the soccer team and the tennis team. I played on the soccer team, but I'm better at tennis. I'm beginning to think that I am better at individual sports than team sports. Hey, maybe I should try golf, the doctor's sport. I've made some new friends this year, too, but I don't have a girlfriend yet.

My father really pushes me to work hard at school. I think he regrets not having the chance, but he grew up in another country and life was very difficult there. He wasn't able to go to university himself, and he tells me to work hard to get the education I can and to offer something back to Canada. He knows that I've learned how valuable medicine can be and that being a doctor would mean helping others. However, it will be difficult for my family to pay for university. My mother died of cancer a few years ago and there are three children to raise. Because I am the oldest, I try to do a lot of the chores, and help with my younger sisters.

I'd like to get a part-time job. It would be good to have some money to buy a few CDs and a new tennis racquet, but, more importantly, I have to save for university! My father says I can't get a job. He says I have to spend my time studying. When I tell him I want to help out, he says it's a father's job to support the family, and my job to go to school. I get an allowance, but I think working will force me to manage my time better and I will learn some work skills. I could work in the summer—that's another possibility to consider.

The disagreement with my father will get worse as I get older. There are not many jobs for people my age, but when I turn 16, it will be a bigger problem. My father just doesn't understand how much it will cost.

What do you think?

1. Why does Adam want to get a part-time job?
2. What thoughts and feelings does Adam's father have about him working?
3. Apply the decision-making process to Adam's problem. What are his alternatives? What would be the consequences of each of them?
4. What do you think Adam should do?
5. How can Adam negotiate with his father?

Right: Health and safety regulations might require that you wear protective clothing on the job.

David didn't live to see his first pay cheque.

He was crushed to death on just his second day working in a bakery. Wasn't supposed to be dangerous, let alone deadly, but it was. You have the right to ask your employer about what's safe and what's not. It's too late for David. But not for you.

HOW SAFE IS YOUR JOB?

If you don't know, you need to. Call us now for more facts on workplace safety.

WSIB CSPAAT 1-888-921-WSIB
www.yworker.com

Left: Awareness programs to inform young workers about their rights and responsibilities in the workplace will help make your work environment safer.

CHECK ✓Points

- ✓ What other aspects of your life might be affected by a part-time job?
- ✓ What are your rights concerning safety in the workplace?
- ✓ What are your responsibilities in the workplace?

You have responsibilities, also, for safety on the job. Do not engage in horseplay on the job because its creates safety problems for you and other workers; for this reason, it is against the law. Report unsafe conditions immediately and take action to clean up hazards such as spills. Wear the protective clothing and equipment that the job requires even if it is ugly. Hairnets, gloves, safety boots, or vests are designed to protect you and others. Make sure you have learned the safety procedures for your job and follow them. Finally, know the location of the fire alarms, fire extinguishers, first-aid kits, and exits, know how to use them, and make sure they are accessible for your safety if something does go wrong.

Volunteering

Volunteering means working without being paid for your time. When you choose to volunteer, you give your time, your energy, and your skills to the organization and to your community. The work that you do goes beyond what the organization can pay for and so is a valuable contribution because without you, it would not be done. Teenagers volunteer for many reasons. Some are required to volunteer as part of their education, but most choose to help an organization because of personal interests.

Volunteering will provide benefits not only for the recipients of your work, but also for you. Unpaid work as a volunteer might provide greater opportunities to experience the type of work you want to do as an adult. For example, you are more likely to find volunteer work at a veterinary clinic than you are to find a paying job. Because your skills are just developing, you are not qualified for most jobs yet, but you could volunteer to help someone else so that you can see for yourself what the job is really like. Volunteering to be a coach, scout leader, or Sunday school teacher, for example, can also provide opportunities for developing leadership skills. Volunteer work listed on your resume tells a future employer that you can manage your time and that you have a positive attitude towards work.

> *The greatest reward for man's toil is not what he gets for it, but what he becomes by it.*
>
> John Ruskin

Who are Canada's volunteers?

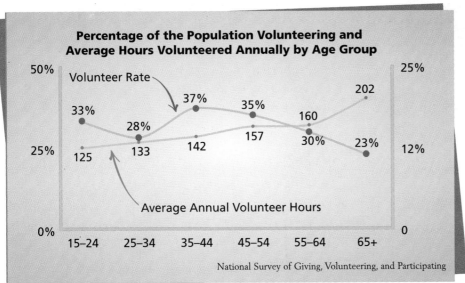

Percentage of the Population Volunteering and Average Hours Volunteered Annually by Age Group

Volunteer Rate

33% 28% 37% 35% 30% 23%

125 133 142 157 160 202

Average Annual Volunteer Hours

15–24 25–34 35–44 45–54 55–64 65+

National Survey of Giving, Volunteering, and Participating

In addition to helping you in your career planning for the future, many teenagers find that there are personal rewards from volunteering. Many kinds of work are interesting, enjoyable, and provide the satisfaction of feeling useful. Working with a variety of people will enhance your relationship skills, and you might form some lasting friendships. The voluntary sector is an important part of Canadian society. As a volunteer, you will contribute to the lives of individuals and families in your community in ways that reflect your interests and values.

How Will You Make Your Contribution?

You have so much to do! There are many things that are part of your lifestyle as a teenager right now. You also know there are developmental tasks that you are accomplishing at this stage in your life as an adolescent. Take advantage of the years that you will be at home before you head off alone on your journey into adulthood. Making the best contribution you can, in a way that suits your family's lifestyle needs, could improve your individual and family lifestyle. Your contribution will also enhance the quality of your relationships with your family, and allow you to become a responsible member of your family and your community. It helps you to become the best you can be. As you read before, there are no guarantees with human behaviour, but the odds are in your favour.

Volunteering to share your skills enables children to experience the same pleasures you did as a child. Your reward is seeing them do it.

Summary

1. The roles of teenagers, males and females, are no longer clearly defined but are determined within families.

2. Strong families manage their resources well, have a clear authority structure, delegate responsibility for performing the functions of families, and are more resilient to challenges.

3. Conflict is common between parents and teenagers because of the teenager's changing interests and values, and the need for more personal power.

4. By taking on more responsibility, teenagers become more independent and can earn the trust and respect of their parents.

5. Sibling rivalry is a common problem but can be reduced if siblings are able to respect their individuality and avoid competition within the family.

6. Caring for families requires many hours of unpaid work. Teenagers can take on some responsibility for this work according to their own skills and abilities.

7. Part-time employment provides additional income but will affect the time available for schoolwork, social activities, and household responsibilities.

8. Paid employment and volunteer work provide benefits for career planning and personal rewards.

In this chapter, you have learned to

- describe strategies for forming effective family relationships;
- describe the various ways in which family members contribute to the family;
- explain the various ways in which teenagers are responsible within families; and
- summarize strategies for negotiating your responsibilities and privileges within a family.

Activities to Demonstrate Your Learning

1. Read a case study about a family that demonstrated resilience in surviving a crisis. Using an organizer, summarize the strengths they demonstrated which made them resilient.

2. Using social science research skills, investigate the number of hours spent on unpaid work, and the nature of the work that is done, for male and female students in your community.

3. As a class, compile a collection of advertisements for products used for household tasks. Determine the values and standards that are reflected in the images and language of the advertisements. Select one of the products and design a counter-ad in which you present values and standards that realistically reflect those of families in your community.

4. Use the Internet to explore the resources available on-line to assist you with a job search offered by Human Resources Development Canada, by your provincial or territorial government, and by other community organizations.

5. Conduct research to determine the nature of volunteer work available for teenagers in your community. Create a directory using database software, or other organizational methods such as a bulletin board, for your student services office.

6. Develop and implement a plan for a class volunteer project to contribute to your community.

7. Reflect on the role that you play in your family. Write a job description that outlines the physical, social, and emotional tasks you perform to contribute to the well-being of your family.

Enrichment

1. If you live in a multicultural community, investigate the expectations placed on male and female teenagers concerning unpaid work, social life (e.g., curfews), and educational goals. Analyse your results to determine whether there are similarities or differences.

2. Interview someone in your faith community about the role of religious or spiritual faith in providing strength for the family. Compare your conclusions with those of students who have investigated other faiths.

Management for Life!

What is the difference between someone who is happy with their life and someone who feels that their life is out of control? What can you do if you need more influence or control over an area of your life? How can you keep your finances in check? What should you do to run your life in a satisfying way?

By the end of this chapter, you will be able to

✳ apply the management process to achieving individual, group, and family goals;

✳ describe the impact of economic, technological, environmental, and health factors on lifestyle decisions;

✳ identify community resources that offer services to individuals and families; and

✳ analyse various ways in which families use resources to perform chief functions.

You will also use these important terms:

community resources	protection
human resources	resources
income	skill
insurance	talent
management process	technology
non-human resources	

What Is Management?

The ability to make effective decisions about your life is important to your happiness. As you have learned, what you want to achieve is usually stated as a goal. You decide on your goals based on your needs and wants. Goals reflect your values and are useful in channelling the direction of your life. For example, you might have the goal of becoming a social worker and take action through your education in order to accomplish that goal. In working towards your goals, you will have to deal with many factors that will affect your achievement. Taking control over factors that affect you is an important step in guiding the direction of your life.

The **management process** helps you to use what you have to get what you want. Within the management process, goal setting will be most effective when you:

Show accountability.
• If you involve others in the process, they may give you the incentive to follow through. Motivation increases when you know that someone else is watching you.

Define the time line.
• Determine whether your goal is long-term, something that will take years to accomplish, or short-term, something that will be done in a few days or weeks.

Write your goals down.
• Seeing your goals on paper helps you stay focussed.

Make goals specific.
• State exactly what you want to have happen. For example, say "I want to get at least 75 percent on my next biology test," rather than just saying that you want to do better.

Remember your goals need to be:

• **S**pecific
• **M**easurable
• **A**ttainable
• **R**ealistic
• **T**imely

Adapted from *The Learner's Edge*

Once your goals are clear, then you must decide how to accomplish them. A key factor in this process is knowing what things you have to help you to achieve the desired results. **Resources** are all the things that you can use to reach your goal. To live a more satisfying life, you need to be able to make the most of all the resources available to you—within yourself, your family, and your community. Resources are valuable to everyone, but some are in limited supply. You will need to make the best possible use of them to meet the needs of yourself and your family.

Two Kinds of Resources

You have two kinds of resources available to you—human and non-human. **Human resources** are those skills, talents, and qualities that come from people. They can include your knowledge, skills, creativity, energy, and time as well as those of your friends and family. ***Non-human resources*** include tangible items that are available to help you such as money, tools, sources of information, or other materials, such as personal possessions.

Human Resources

Your personal set of human resources may be the easiest and most useful resources to draw on to achieve your goals. You have these qualities with you at all times. As you examine the different types of human resources, consider what personal qualities and characteristics you have to help you in the management process.

> *Do what you can, with what you have, where you are.*
>
> Theodore Roosevelt

Knowledge

There are many different kinds of knowledge. Your knowledge includes everything you have observed, learned, and remembered up to now. You are able to increase your knowledge in any area that interests you. Another very important type of knowledge is knowing how and where to access information. If you do not have knowledge about a specific topic, it is important to know how to find it. When you have the facts and the understanding to do something, you have very useful knowledge. Knowing how to do things, for example, to speak French or to fix things around the house, can benefit you and your family. You accumulate knowledge throughout your life—learning is a lifelong process.

Where do I get knowledge and information?

- **Ask a friend.**
- **Ask a teacher or parent.**
- **Go to the library.**
- **Search the Internet.**
- **Take a course.**
- **Attend a workshop.**
- **Ask an expert.**

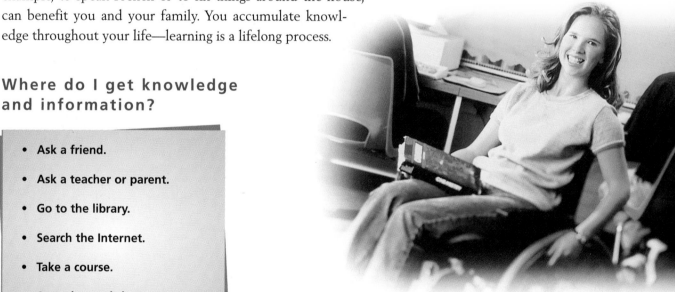

Knowledge is one of the most important resources you have. In school, you develop the skills necessary for learning. Various courses provide you with the knowledge you will need to make decisions in your life.

Skills and Talents

Skills and talents are the abilities to do something well. A ***skill*** is an ability that you have learned, such as riding a bicycle or using a computer. Every skill that you acquire may become a resource for you some time in the future.

Left: Learning a variety of skills provides a wide variety of resources for you to use.

Above Right: A natural talent may create the ability to gain material resources. A talent for designing clothes could lead to a successful career in the fashion industry.

Working in community or school clubs gives you the chance to gain leadership experience. How can working in small groups give all students a chance to enhance leadership skills?

Skills for communication, reading, and math are basic resources you acquire in school. For example, the keyboarding skills that you learn in school may help you to perform your work in a future job. A *talent*, on the other hand, is an ability that comes naturally, such as the ability to sing in perfect pitch. Talents can also be used to achieve your goals. For example, a talented artist might make new friends, learn new skills, and earn recognition by painting the sets for the school play. By investing your time and energy in training and practice, you can enhance your talents and skills.

Personal Qualities

Your personal qualities and personality traits may be useful as resources. If you have emotional resources such as the ability to empathize with others or an outgoing and friendly nature, you possess the personal resources necessary to motivate others. Having a positive attitude and a drive to succeed are factors that will contribute to your own motivation. Trusting yourself will allow your creativity to shine through. Creativity and imagination are often powerful resources in helping to create unique solutions for difficult problems. Personality traits that others trust and value combined with the ability to bring out the best in people produce the right mix for leadership. Leadership qualities can be used to further your personal goals or the goals of a group.

Energy

Energy is your ability to be active and accomplish tasks. Your energy level is determined by many factors. Physically, your metabolism, what you eat, and how much rest you get affect your energy level. Having good health is an important component in your ability to use energy. Your attitude and general outlook on life affects your emotional energy level.

Left: Your attitude and your outlook can really affect your energy level. Maintaining healthy lifestyle habits contributes to high physical energy.

Knowing about time-wasters and time-savers can help you use your time better.

Time

Time is a limited human resource since everyone has only 24 hours each day. Whether you use your time effectively depends on your ability to make decisions about managing it. In our society, time is viewed as a scarce resource. There are so many demands made on your time that, like most Canadians, you probably feel "time-stress" sometimes. It is important to learn about time-savers and time-wasters in order to make time work for you.

Other People

Your family and friends possess all of the same types of human resources as you do. They have knowledge, skills, and personal qualities that may help you with a problem. They may also have the time and energy to help you get a job done. When family members work together, they can provide excellent resources for each other. Teachers are excellent resources to help you gain the knowledge that you need. Experts and specialists in many fields provide you with knowledge about areas that you are unfamiliar with.

CASE STUDY

Mario's Story

Here is my example of how using the resources that are available can allow you to achieve your goal.

We had to work as a group to design a car that would run on a rubber band. Our teacher, Mr. Schnalzer, said we could use any materials we wanted from those that he provided.

At first I thought the assignment was impossible, but once we got started, we discovered that we had many other resources that enabled us to be successful. We were learning how cars are designed, so we were expected to apply that knowledge. We had also learned how to use the tools that we needed.

Brett became the leader of the group. He got us excited about making the best car. We went over to his house one day after school and sketched several ideas. Harvey could draw well so he helped us to come up with a design that looked great, and then he did most of the painting when we finished the car.

Steve helped us all to understand how the rubber band had to provide the energy to drive the car. Mr. Schnalzer helped us figure out a few technical details at lunchtime, and then I kept working on the mechanics until we had it adjusted perfectly to make it the fastest car. We even finished the car on time.

I think we got carried away with the final appearance of the car but, in the end, that helped us get a top mark. We got a great mark on our project in technological studies because of all the resources we used.

What do you think?

1. What resources enabled Mario and his group to be successful?
2. Which of their resources would everyone in the class have?
3. Which resources would only some people have?
4. What choices did they make about using the available resources?

CHECK Points

✓ What is the difference between a skill and a talent?
✓ What factors can affect your energy level?
✓ How can you acquire knowledge?
✓ List the talents that you have. How are you able to use them?
✓ What personal qualities do you possess that would be useful as a resource?

Non-Human Resources

You are more familiar with non-human resources. In order to achieve many of your goals, you will use possessions of your own or those that are available in your community. You already have many possessions, and you can obtain others that you need if you can afford to buy them. There are resources in your community and in the environment.

Material Resources

Tangible objects that you can use to accomplish your goals are material resources. Possessions are things that you own that can perform some kind of function. Your bed provides a comfortable sleeping surface, your clothes keep you warm, and your television set can inform or entertain you. Some possessions, such as houses, furniture, and cars can last a long time. Others, such as paper, glue, and food, are used up quickly and need frequent replacement. Possessions are used to create other resources. Tools can be used to fix a car or to utilize a skill such as carpentry. Possessions can also be exchanged for other possessions or for money.

Money

Long ago, people did not need money. Using the barter system, people exchanged what they had for what they needed. For example, a farmer might give a blacksmith several bags of corn in exchange for having shoes put on his horse. Simple bartering was replaced by money as society became more complex. In modern societies, people receive money as *income*, and receive goods and services in exchange for money. Today money is very important as it provides you with the means to buy the other resources that you need and want to achieve your goals.

Above: Money is a resource that can be used to acquire other resources. This teen will use the bicycle to earn money by delivering flyers.

Left: When possessions are no longer useful to family members, they can be sold. Others see value in these possessions and will pay money for them.

Community Resources

Every community has organizations and facilities that help you to enhance the quality of your life. Schools, libraries, and museums provide information and learning opportunities to support your intellectual growth. Community centres, parks, and theatres provide opportunities for entertainment. Hospitals and medical clinics offer services for health care to maintain your physical well-being. *Community resources* come in many forms and serve many functions. Some community resources can meet many needs. Community centres provide a source of information about community events as well as recreational programs and a day care.

Houses of worship are important resources that support your spiritual needs. Churches, cathedrals, temples, mosques, and synagogues serve their congregations. A congregation is a group of individuals who have chosen to form a religious community. There are many religious communities within our society. Each family chooses whether or not they will be part of a religious community and, if so, which religious affiliation will best suit their needs. A house of worship provides a welcoming environment for spiritual expression.

Businesses within the community also provide many valuable resources. They allow you to purchase products and services to meet the needs of you and your family. Businesses also provide you with the opportunity for employment. Through part-time or summer employment you will be able to earn money, learn new skills, and gain experience. Businesses often provide support to other community resources as well. They may support a local little league team or sponsor a community fundraising event.

Natural Resources

The earth provides all of us with many resources. We all need and use natural resources such as air, water, soil, animals, plants, and minerals to meet many of our basic needs. Humans alter many of these natural resources to create other material resources. Trees are used to make paper; some plants are used for food. Water and coal are both used to generate electricity, which is used as our most important source of energy. Our natural environment also contributes to our emotional well-being. The beauty within the natural environment can provide us with a sense of peace and tranquillity.

Natural Resources

Water	electricity, transportation, drinking, washing
Air	breathing, cooling motors
Plants	trees for paper, wood; plants for food; cotton for clothing
Animals	for food, sheep for clothing, horses for work
Minerals	coal for heat, silver for money

By using natural resources on a large scale, humans have altered the natural environment tremendously. Large forest areas have been eliminated, the course of rivers has been altered, and the number and types of wildlife in many areas have been reduced. Sometimes the result of humans' use of natural resources is pollution. Pollution is the addition of dirt and toxins to the air and water. It is important that people conserve, or save natural resources so that the earth has a sustainable future. Pollution can cause harm to animals including illness in human beings. For our natural environment to be healthy, we need to learn to use our natural resources wisely.

career Link

What is an environmental analyst?
An environmental analyst is skilled at analysis of the environment.

What does an environmental analyst do?
An environmental analyst conducts basic and applied research to extend knowledge of living organisms, to manage natural resources, and to make recommendations about the use of natural resources. They conduct ecological and environmental impact studies and prepare reports and plans for management of renewable resources. For example, they might investigate the impact of putting a new residential development into an agricultural area. They report their findings to government agencies and to the public so that appropriate decisions can be made about resource use.

Where do environmental analysts work?
While most analysts are employed by federal, provincial, and municipal governments, many work on a freelance basis. Business and industry in areas such as mining, forestry, and agriculture may also employ them.

The beauty of nature can provide us with the inspiration for creativity. It is important for us to use natural resources wisely in order to conserve them for the future.

Technology

Technology is a human-made resource that combines scientific knowledge with natural resources to create tools that extend your abilities beyond your own physical or mental power. Machinery, such as washing machines and bread machines, allow us to work more efficiently than when these tasks were done manually. Telephones let us talk to others who are far away from us. Medical technology, such as pacemakers to regulate heartbeat, dialysis machines to support kidney function, and lasers for precision surgery, has contributed to the general health and well-being of individuals and families. Computers provide a wide variety of information sources in one location. Just as machinery helped us to overcome the hardships of physical work during the industrial revolution, today's computer systems and imaging processes are helping to facilitate analytical work in our fast-paced information age.

Technological advancements such as an "at-home" blood sugar monitor allow individuals to participate in the monitoring of their own health.

Technology has given us many helpful household tools and machines that help people look after today's family.

CHECK ✓ Points

✓ Identify your most important personal material possessions. Describe how they help to meet your needs.

✓ Identify the natural resources you and your family use on a daily basis.

✓ Why is it important to conserve natural resources?

✓ Discuss how your family uses technology in your home.

Reflections and Connections

1. Identify the skills and talents that your family members possess. How can they be used as resources within the family?

2. What are the similarities and differences between material and natural resources?

3. Which of the natural resources used by your family are available in Canada?

4. How are community resources paid for? Would it be better to have a "user pay" system?

5. In what ways can you increase the community resources available in your community?

6. How are technological systems used in your community?

7. In your opinion, does all technology benefit families?

Using Your Resources

Resources that could be used to help you achieve your goals surround you. Good management is a multistage process. First, you need to state your goal clearly in order to meet your needs successfully. Once you have identified your needs, identify what resources are available for you to use. Then, apply the decision-making or problem-solving process to consider your alternatives. What resources you could use will determine which of the alternatives are realistic possibilities.

We cannot live for more than a few minutes without air, a few days without water, a few weeks without food. Yet we continue to destroy the very things that keep us alive and in the process rob our children of a future.

Anita Gordon and David Suzuki

Identifying Your Resources

Most resources are scarce, which means they are in limited supply. You need to use each resource in the way that will be of most benefit to you. In any given situation, there are many resources available to you.

By making the jacket herself, Lorraine would use her skills, time, and energy. By buying the jacket, she would be using the material resource of money as well as the community resource of a clothing store. By borrowing a jacket from her cousin or by asking her parents for one for her birthday, she would be using the resources of other people. In every situation, there are many resources available.

Lorraine's Resources

Lorraine needs a waterproof jacket for a camping trip next month. There are several ways of achieving her goal:

1. **Make the jacket herself.**
2. **Buy the jacket with money that she has earned.**
3. **Borrow a jacket from her cousin.**
4. **Ask her parents to give her a jacket for her birthday in three weeks.**

Sharing Resources

Most resources can be shared. Families share resources on a daily basis. All members of the family share living space and furniture. Family members often share time and skills when preparing meals and looking after other family needs. You share your abilities and attitudes with your family every day.

An example of shared human resources is a team at school. Each person shares their athletic skills, their energy, and their team spirit. This type of sharing can be enjoyable to everyone on the team. When this sharing takes place on a regular basis, the team's skills are improved and all members feel successful.

Everyone in the world shares natural resources. Since air, water, and land are limited resources, it is important for all of us to conserve these resources. Community resources such as water treatment plants and conservation authorities help to conserve natural resources in your community.

Exchanging and Substituting Resources

As the example of Lorraine's jacket demonstrated, any number of resources can be used to meet a specific need. Before the use of money became popular, people would regularly exchange one resource for another. A tailor would make clothing for a family, in exchange for food that the family produced on their farm. The practice of exchange is still used today; however, money is the most common medium of exchange. Today you exchange money for food and other products.

Sometimes you might exchange resources within your family. For example, you might do some work for your father in exchange for getting the use of his car. Also important, but less obvious is the exchange of personal resources such as emotional support and empathy within the family. Family members exchange caring, comfort, love, and loyalty. It is vital to receive these resources from family members, especially when you feel sad, disheartened, worried, or upset. Family members all go through tough times. Exchanging emotional resources in times of crisis or sadness will help each member to cope and support each other.

Money is exchanged for goods and services. You provide a service to your employer for money. You then exchange that money for the goods and services that you need and want.

The Management Process

Once you have determined your goal and all of the resources available for a particular situation, you are ready to go ahead with the management process.

The management process closely parallels the decision-making process; however, the emphasis is on the use of resources. Management issues focus on how to use your resources to achieve a desirable outcome.

For any given goal ask yourself:

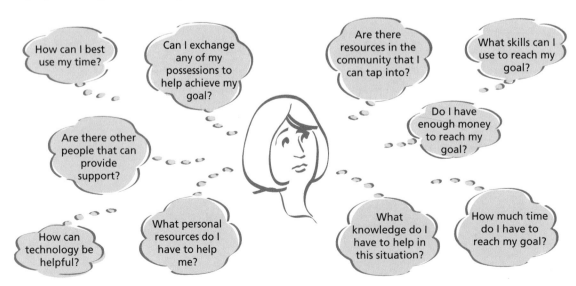

How can I best use my time?

Can I exchange any of my possessions to help achieve my goal?

Are there resources in the community that I can tap into?

What skills can I use to reach my goal?

Are there other people that can provide support?

Do I have enough money to reach my goal?

How can technology be helpful?

What personal resources do I have to help me?

What knowledge do I have to help in this situation?

How much time do I have to reach my goal?

With each of the possibilities you may be able to identify pros and cons. Recall that you will need to consider your values when deciding on positive and negative aspects of each resource use. For example, you might decide that your time is more valuable than your money, or that human resources are more important than material ones. This would direct you to making a purchase rather than taking the time to make something yourself.

CHECK Points

✓ Today we often think only of computers when referring to technology. Describe five other ways in which technology can be used as a resource.

✓ Many resources can be shared. Identify and describe three human and three non-human resources that can be shared.

✓ What are the similarities and differences between the management process and the decision-making process?

Decision-Making Process

- Identify the issue.
- Consider your standards.
- Identify alternatives.
- Predict consequences.
- Estimate probability.
- Take responsibility.
- Carry out your plan of action.
- Evaluate results.

Management Process

- Establish a SMART Goal.
- Identify your available resources.
- Consider the best use of your resources.
- Prepare a resource plan (how to acquire, substitute, or share resources).
- Implement the plan.
- Evaluate the plan. Did you reach your goal?

Reflections and Connections

1. Think about the last major or important purchase that you made. List all of the resources that were involved. Do not forget to consider natural resources.

2. What resources did your family use the last time they needed a major item such as a car or a washing machine? Were you involved in the process?

3. Describe some ways that you share resources within your family.

Protecting Your Resources

An important consideration when evaluating your resources is how safe they are. If you are in good health and take care of yourself, you will likely continue to have high energy as a resource. However, if you engage in activities that put your health at risk, then you may not be able to depend on your personal energy as a consistent resource.

Consider the following situations:

- Jerome and his family had gone for the weekend to visit friends in another town. When they arrived home on Sunday evening, they discovered that their house had been broken into. Thieves had taken all of their valuable possessions. They lost their TV and stereo as well as a lot of precious heirloom jewellery. Jerome's father was particularly upset because his computer was stolen. He runs his business from home and all his client records were gone.

- Suzanne came home from school one day to find her mother had come home early from work. She was crying and very upset. She told Suzanne and her sister that their father had been in a car accident that day. He had been pronounced dead when the ambulance had arrived at the hospital. Suzanne and her family had lost the love and support of their father. His death also made them a single-parent family with only one income.

Resources such as your possessions, your money, your health, or even a family member's life can be at risk. Since many resources are critical to caring for family members, they often require some *protection*.

In order to protect their resources, many families take out various types of insurance. *Insurance* is a form of protection that can be purchased from insurance companies. In exchange for agreeing to pay a set amount on a regular basis, a family can be assured of receiving money in the event of a loss. An insurance policy, which is a contract between you and the insurance company, can be purchased to compensate for the loss of health, life, or material possessions of you and your family. Non-human resources that are commonly insured include houses and their contents, cars, and jewellery.

Theft can cause the loss of many family resources. Money and other valuables as well as important technological resources may be lost.

CHECK Points

✓ Why do families purchase insurance?
✓ What kinds of insurance are available?
✓ Describe how insurance can be a substitution of resources.

Summary

1. The management process involves making decisions about using your resources to accomplish your goals.

2. Human resources are qualities that come from people, such as skills, talent, knowledge, and creativity. Non-human resources are tangible items such as money and tools.

3. Human resources are personal and can be drawn upon at any time. Non-human resources are often limited and can be used up.

4. Many community resources such as libraries provide both human and non-human resources. Librarians provide the knowledge and skills that people need to help them get information while the book collection provides tangible resources for community members to use.

5. Since resources are valuable, they can be used, preserved, substituted, and shared.

6. Resources help families to care for their members. Resources sometimes require protection in the form of insurance to assure the security of families.

In this chapter, you have learned to

- apply the management process to achieving individual, group, and family goals;
- describe the impact of economic, technological, environmental, and health factors on lifestyle decisions;
- identify community resources that offer services to individuals and families; and
- analyse various ways in which families use resources to perform their functions.

Activities to Demonstrate Your Learning

1. Make a list of your personal resources. Categorize your resources into human and non-human resources. Identify the ones that you might develop further and list ways in which you might do so.

2. Using social science research skills, conduct a survey of the human resources students have acquired in school. Make a bar graph of your results.

3. Identify a long-term goal that you have. Using the management process, outline what resources you could use to achieve your goal. Suggest a plan for acquiring the necessary resources.

4. Conduct an interview with an adult to find out what resources they protect through insurance and why they protect those resources.

Enrichment

1. Contact an environmental group in your community. Find out what concerns they have about the use of natural resources in your area. Prepare a report about the long-term effects on your community.

2. Identify a personal goal that you have. Make a list of all of the possible resources that are available to you to reach your goal. Using the decision-making process, make a plan of how you can use your resources to achieve your goal. Make sure you consider resource substitution and resource sharing in your plan.

3. Search the Internet for resources that can help you to increase your knowledge in an area of interest to you. Prepare a report for the class about how you conducted your search and the results of your investigations.

13

Managing Your Time

Time has always been a central focus for people. We all have the same amount of time each day, yet we do not feel the same way about the amount of time we have. Some people seem to accomplish a lot each day, while others never seem to have enough time. Do you often feel rushed? Are you able to meet all of your deadlines, or are projects often late? Do you have enough time to do the things that you want to do? How you use your time is important to the quality of your life.

By the end of this chapter, you will be able to

✳ apply strategies for managing your time to achieve individual, family, and group goals;

✳ describe strategies for acquiring, increasing, and substituting your resources; and

✳ describe the impact of economic, social, technological, environmental, and health factors on lifestyle decisions.

You will also use these important terms:

deadline	priorities
dovetailing	procrastination
interruptions	schedule
multitasking	time management

Why Manage Time?

You have the same amount of time as everyone else. How often have you found yourself saying "I needed more time" or "I ran out of time"? Sometimes others may ask you to do things with your time. For example, your teacher may ask you to complete a project within two weeks. You also have things that you want to do for yourself. Some activities have time limits, such as shopping for something to wear to the party on the weekend. No matter what the activity or goal, if you want to accomplish it, you need to use some of your time.

In some cases, there are many things to do within a single time frame. While you have to work on that project the teacher assigned, you still have to get all your other homework done, play on the school basketball team, contribute to work at home, do your baby-sitting job, watch your favourite television programs, talk to your friends on the phone, and socialize. How can you possibly get it all done? The answer is ***time management***! By using time-management strategies, you will be able to accomplish your goals in a timely manner and ensure that all the things you want to do get done.

Time Management and Balance in Your Life

An old saying states that "All work and no play makes Jack a dull boy." On the other hand, too much play does not allow for much work to be done. It is important that you have a balance of work and recreational activities in your life. You have certain obligations to yourself and to your family such as getting good grades in school and helping out at home. You also have obligations to your friends, such as spending time with them to maintain good relationships and to enjoy each other's company. Whenever work and recreational activities are not in balance, the quality of your life and your relationships may suffer.

How do you decide how much time to spend on certain activities? The first step is to review your values and goals. You must always keep in mind what is most important to you when you consider time management. It is also important to keep in mind the difference between your needs and wants. For example, you may want to play video games every day, but you need to pass Grade 10. Some activities must be given a higher priority.

Identifying Time Categories

In order to begin identifying the important ways you use time in your life, you can group your activities into categories.

Time Categories

Looking After Me—Personal Care

Personal care uses more of your time than any other activity. Personal care includes eating, sleeping, dressing, bathing, combing your hair, and other personal hygiene activities. Many personal care activities are important to maintain good health. For example, you need adequate sleep to function to your peak potential.

You may spend almost 12 hours a day in eating, sleeping, grooming, and other personal care activities.

Taking Care of Business—Work

Productive activities take up a large part of your day. Attending school, studying, and doing homework are your primary tasks in this category. You have to do your share of the household duties. You may also have part-time, paid, or volunteer work as part of your activities.

You may spend from six to nine hours each day in work-related activities such as school, homework, house-work, and a part-time job.

Time to Kick Back—Leisure

Having fun and relaxing needs to be a regular part of your day. You may like to read or watch television to relax. Playing softball with your friends might be your idea of having fun. You may have a hobby, such as model building that you find both relaxing and fun.

You may find reading novels a good way to relax.

Connecting with Others—Relationships

Spending time with others can greatly enhance your life. Your family and friends can also be important resources for you. They provide you with caring and support and in turn, you need to spend time providing care and support for them.

Spending time helping friends to work out problems can strengthen your relationships with them.

What They Do with Their Time

The average time spent per week on various activities among men and women aged 35 to 44

	Hours per week	
	Males	Females
Total paid work and related activities	43.5	26.6
Paid work	39.2	23.8
Activities related to paid work	0.7	0.0
Commuting	3.5	2.1
Household work	19.6	35.0
Civic and voluntary activity	2.1	2.8
Night sleep	54.6	56.0
Meals and other personal activities	14.0	15.4
Socializing	10.5	11.9
Television	13.3	11.2
Education and reading	2.8	3.5
Other leisure	7.7	6.3

Statistics Canada, *General Social Survey, Cycle 12*: Time Use, 1998

Assuming that you will want to balance these time categories, you will need to identify your values and set goals for each area. For example, in the personal care category, you may set a goal of getting at least eight hours of sleep each night. Decisions you make about one category of time use will have an impact on the other areas. For example, by deciding to have eight hours of sleep each day, you will not have the time to watch late-night television.

Some activities can help you to reach goals in more than one area. Here are some examples:

1. Eating at a restaurant with a friend looks after your personal care need of eating and also allows you to spend time on your relationship with your friend.
2. Helping your dad with the garden enables you to contribute to the work at home as well as spend some time "bonding" with your father.
3. Reading and making study notes for chemistry during a camping weekend allows you to meet some of your school-related goals while still participating in a leisure activity.

Many times, opportunities are available to do more than one thing at a time. This process is called ***dovetailing***. When you dovetail activities, you are using "down-time" within one activity to accomplish another. For example, if you have a doctor's appointment after school, you can spend the time in the waiting room reading the chapter you were assigned for homework. By doing this, you free up time later in the evening to do other things.

Check Points

✓ Why is time management important?

✓ What factors affect the way time is used?

✓ How can "dovetailing" save time?

Reflections and Connections

1. Is there a balance in your personal use of time? Why or why not?
2. List all of the tasks that you need to do in the "personal care" and "taking care of business" categories. Do you have enough time to get them all done?
3. Describe a time when you really planned for quality time with an important person in your life.

How Do You Want to Use Your Time?

Once you have determined your values and goals in the four general time categories in your life, you need to make a plan for each area. Creating a plan for your time will allow you to make decisions to support that plan.

The first step in the plan is to make a list of all the things that you need and want to do. This can be a simple "to do" list. By sorting these tasks by category, you will immediately see whether you are aiming towards a balanced lifestyle. If all the tasks on the list are school- and work-related, perhaps you need to look more closely at time spent in leisure and relationship activities. If all your tasks revolve around your friends, perhaps your grades and personal care needs are not getting the attention that they need.

I must govern the clock, not be governed by it.

Golda Meir

In creating a plan, you need to decide on your **priorities**, the most important things that need to be done. Your priorities may be different from those of your friends. To decide on your priorities, you can examine which tasks and related goals are most important to you. The importance of each task or goal will be based on your personal values and standards. In each time category, you could assign a priority to each task.

Priority 1 items: things that must be done

Priority 2 items: things that should be done

Priority 3 items: things that you hope to do

Some items have deadlines. A **deadline** is a specific time by which a task must be done. Often these are the first priority items. School assignments usually have deadlines. Many other tasks also have deadlines; for example, if you want to get a gift to take to your friend's birthday party, you have a deadline. Although you may end up giving your friend the gift after the party, the deadline to meet your goal was the time that you arrived at the party.

Your to do list for work-related and personal activities might look like this:

Things to Do	Deadline for Completion	Priority
Read Chapter 4 in History text	Monday	1
Return library books	Friday	2
Clean room	Saturday	3
Finish Science project	Next Wednesday	2
Return Jodi's jacket	Before Friday	2
Complete summer job application	Next month	2
Do Math homework	Wednesday	1
Re-organize closet	Next month	3
Send Maureen a birthday card	By next Tuesday	1

Getting Yourself Organized

Knowing what you need to do and what your priorities are is important. Having a method to implement your to do list is the key to a balanced and satisfying use of your time. Making a plan for your time is much like planning for your goals. You need to plan in the short term, in the long term, and also in the interim period. Your short-term time plan would include what you need to do on a daily and weekly basis. Your long-term time plan would look at tasks and goals that need to be accomplished within a one- to five-year time span. Interim plans would be for the next few months.

Time-Management Tools

There are several time-management tools that can help you get organized. You may have a school planner or other personal planner to help you with this process. The school planner is an example of a calendar. At the beginning of the school year, this type of planner tells you major time frames, such as when exam times and school holidays will occur. You can use this planner as an interim organizer in many ways. You can use it to record homework assignments and test days. It can also be a reminder for doctor's appointments, birthdays, and other special events.

A **schedule** is a way of planning time on a daily basis. If you are a person who likes to plan in detail, a schedule allows you to precisely allocate time to the various tasks in your life. Many business people use a day-at-a-glance planner to schedule their appointments. Although schedules are helpful for some people, they can be time-consuming to manage. Sometimes a to do list is all that is necessary. A to do list that groups activities together can be made up for a specific day, event, or task. The items listed on a to do list can be incorporated into a daily schedule or a calendar.

Daily Schedule

	To Do		
6:30	Shower and get dressed		
7:00	Breakfast with Mom	Why?	Tuesdays are the only days that Mom goes to work late.
7:45	Leave for school		
8:00	Choir practice	Why?	Practices are Tuesday and Thursday.
8:45	Home room		
9:00	English	Why?	Writing down the classes helps to keep a tumbling timetable straight.
10:20	Math		
11:45	Lunch with Leslie	Why?	Cannot forget to meet her!
12:00	Biology Lab		
1:20	Individual & Family Living		
3:00	Tutorial	Why?	I need help with Math.
3:30	Swim practice		
5:00	home	Why?	Help with dinner/cleaning.
5:45	Dinner		
6:30	Relax and watch TV	Why?	Sometimes I record the show so I will not forget that my favourite one is on.
7:00	Homework		
8:00	Science project	Why?	If I do not write this in, I might be tempted to put it off!
9:00	Call Lorraine	Why?	Plan for the weekend.
10:00	Relax; get ready for bed		

monthly calendar

Sunday	Monday	Tuesday	Wednesday	Thursday	Friday	Saturday
1	2	3	4	5	6	7
Nana visits					complete science project	pick up skates
					13	14
sister's birthday			dentist 4:30			concert tickets go on sale
15					20	21
					geo test	basketball game
22					27	28
29	30	31				

Birthday Party To Do List
make up invitation list
buy invitations
send invitations
buy decorations
buy present
decide on food
buy food

shopping To Do List
snack for after the game
new sneakers
deodorant and toothpaste
computer magazine
Back street Boys CD for gift

CHECK **Points**

✓ Why is it important to establish priorities when managing your time?

✓ Are deadlines only for work-related activities? Explain.

✓ Describe the similarities and differences among a calendar, a schedule, and a to do list.

A combination of these time-management tools can help you to organize your time to your satisfaction. You need to experiment with these tools to see what works best for you. Once you have the tools available to use, you can make time planning one of the first steps in completing any project.

Reflections and **Connections**

1. How do you set priorities for using your time? Give an example of how one of your values affects how you use your time.

2. Look in your school calendar or planner. List the types of activities you have recorded there. Do you use it for personal as well as school-related tasks?

3. Choose a task that you need to complete in the near future. Identify all of the things you need to do to accomplish this task by making a detailed to do list.

Time-Management Troubleshooting

Even with the best-laid plans, time management can run into difficulty. Some trouble spots can be avoided; others can be planned for.

Time-Wasters

Have you ever had days where you feel that you have not accomplished a thing? Reflect on one of those days and assess your use of time. Did you spend too much time on the phone? Did you allow yourself to be distracted by an unimportant activity, such as straightening out the contents of a drawer? Did you spend more time doing a leisure activity than you had planned? Time-wasters are different for everyone. The key to avoiding time-wasters is recognizing which ones occur frequently in your life. Once you have identified your personal time-wasters, be vigilant and limit the time you spend on those activities. Remember that you are in control of what you do with your time.

Interruptions

Have you been in the situation where you are just getting into doing your homework and a friend calls? You tell your friend you cannot talk right now and get back to your work as quickly as possible. You just get going again, when another friend calls. This time, you do not get off the phone so quickly. By the time you get back to your work, you have lost the enthusiasm you had for the task that you had when you started. *Interruptions* break your concentration and also waste precious study time. If homework time is a priority, then you can avoid interruptions in a number of ways. You could tell your friends the times that you study and ask them not to call during that time. You could also ask your family members to take messages from anyone who calls during your study time and return the calls after you are finished your homework.

Technology

Use technology to help you with interruptions.

Telephone solutions	call answer/call waiting
Computer solutions	e-mail (respond to messages when convenient, communicate with many people at once)

Procrastination

Procrastination, or putting things off, is a common reaction when we have to spend time doing things that we do not like or that we find difficult. When procrastination causes us to fail at important tasks, problems arise. A common example is leaving a large project until the night before it is due. It is almost impossible to be successful in these situations. Using realistic schedules can be helpful. Anticipating possible "disasters," such as the one book you need being out of the library, or Internet service being disrupted just when you need it, will help you plan enough time to try again the next day. Scheduling

Procrastination is the thief of time.

Edward Young

unpleasant or difficult tasks first can also help. Often tackling difficult tasks early and in small manageable chunks, makes them less difficult and can even provide a sense of success.

Over-Commitment

Saying yes to too many people and projects can leave you with more tasks to do than is possible for one person to accomplish. Examine your priorities. Make commitments only to those activities that you truly value and you can realistically build into your schedule. When you overload yourself with too many activities you may become stressed.

Unexpected Events

Unexpected events can sometimes become time-wasters. Perhaps the bus you are taking to school has a flat tire, or the planning session you had scheduled with the yearbook committee had to be postponed due to a change in the school schedule. In each case, you are unable to accomplish the planned task in the time allotted. The key in this situation is flexibility. It is always a good idea to leave more time than you think you will need. This will compensate for potential delays. When things do not go on schedule, use the time to complete another task, rather than wasting time. This strategy will free up time later to complete the tasks that were delayed earlier.

Have you ever been unable to finish a project on time because your printer ran out of ink at the last minute? Have you gone to the movies and discovered that the movie you went to see is not playing anymore? Have you planned to make dinner and discovered that there are no potatoes? Although these may be surprises for you, all of these events can be anticipated. Planning your time sometimes requires that you check that the resources you need are on hand. Keeping school supplies, personal care products, and staple food products on hand costs no more but can help you manage your time better.

Reflections and Connections

1. Make a list of personal time-wasters. Describe how you might avoid these time-wasters. Show the list to a friend and see if they have other helpful suggestions.
2. What types of things do you procrastinate about? Make a list of specific strategies you could use to avoid procrastination.
3. Identify the things for which you have made time commitments. Describe the values that have motivated you to make those time commitments. Have you made any time commitments that do not reflect your values? If so, what do you want to do about them?

Managing Your Learning Time

The primary work-related task in your life at this time is to be successful in school. You spend many hours of your day in school. You also need to spend time in the evenings and on weekends doing homework, studying, and completing projects. You also want to have a life! Many students value the money they earn from part-time jobs over getting high marks. This is a temporary goal, however. Making schoolwork a priority may provide lifelong benefits. There are many skills and strategies that you can learn to make the most of your learning time.

> *To choose time is to make time.*
>
> Francis Bacon

Note Making

Taking useful notes is an art. Well-constructed notes will help you to remember information as well as organize your learning. There are many note-making strategies. Do not stick with one style or strategy out of habit—you need to match the strategy to your individual learning style and to the learning task at hand. Remember that notes must be legible and organized to be of any value.

Notes on Notes

- Well-organized notes are a great study tool.
- Learn a variety of note-making strategies.
- Take notes in point form in your own words.
- Double-space your notes, if necessary, so they are easyto read.
- Use brightly coloured highlighters to emphasize key points.
- Date your notes to help keep them organized.

Participating in Class

Paying attention in class goes a long way in the learning process. Your teachers are your best resource for passing your courses. Do not let others distract you from the lesson. By using good active listening skills, and by attempting to answer all questions, you can make sure that you are getting the information. Asking questions enables you to clarify the information and extend your understanding. You can also get important feedback about possible misunderstandings that could waste your time.

Active listening is important when learning new information. By asking questions during class, you are checking your understanding.

Reinforce Your Learning

Do your homework as soon as you can. That way the new information is still fresh in your mind and you will be able to apply it more effectively. You can avoid procrastination problems by tackling the hardest tasks first. You may be setting yourself up for failure by trying to tackle difficult areas at the end of a study session when you are tired.

The Curve of Forgetting

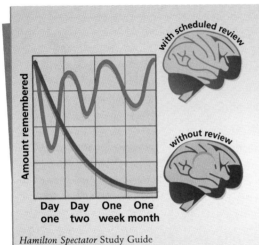

with scheduled review

without review

Hamilton Spectator Study Guide

The diagram shows the impact of repeated review of information. If you read information only once, you are quite likely to forget it as time goes by. When information is revisited many times, the process engages your long-term memory. By using your long-term memory you are more likely to remember information.

Jordanna and Eadonn are working on the chapter on Time Management in their Family Studies text. They both know that it is a very important chapter that will be included on the test at the end of the term in three weeks. Jordanna takes about 30 minutes to read the text and then spends another 30 minutes reviewing the information to ensure that she understands everything.

Eadonn also reads the chapter in about 30 minutes, but does not do any more work on it that day. The following week he spends 10 minutes reviewing the chapter. About two weeks later, he spends another 10 minutes looking over the material.

According to The Curve of Forgetting, who will remember more information for the test?

Reward Yourself

Give yourself a pat on the back—even for small successes. Often others are unaware of your achievements. It is important to acknowledge each small step in your learning progress. Whether you choose to cross items off your list, make charts, or paper your room with completed and marked assignments, keep track of your achievements. By acknowledging learning, you can motivate yourself to keep going.

Managing Your Studying Environment

- Make sure you have adequate lighting.
- Make sure you have everything you need before you start to avoid interruptions.
- Make sure you have good air circulation. If the air is stuffy and you are feeling drowsy, open a window.
- Take short breaks to avoid boredom and maintain a fresh outlook.
- Keep some food and water handy in case you need some refreshment.
- If your home is too noisy or busy, find another place to study.

CHECK ✓Points

✓ Describe some strategies you could use to avoid procrastination with schoolwork.

✓ Why are notes a helpful study tool?

✓ What can you do to avoid interruptions while studying?

Time Management and Resource Substitution

When you have established your priorities you need to ensure that there is enough time for you to fulfil your goals. If there are more tasks than can be accomplished in a certain time frame, you may be able to substitute other resources so that all of your goals will be met. Although everyone has the same 24 hours to use each day, you have varying amounts of other resources that you can use to substitute for time. For example, you might choose to save the time taken in food preparation and kitchen clean-up by going to a restaurant. In doing so, you are substituting money for time. Using some time to tutor a friend for the Math test might be a good trade-off if your friend will use her keyboarding skills to type your assignment faster and better than you could. *Multitasking* can often be accomplished with the use of technology. Be creative in finding ways to save time by using other resources.

career Link

What is a management consultant?

A management consultant is a person who analyses and provides advice on management methods.

What does a management consultant do?

Management consultants perform many duties. All of their duties require decisions about how time will be used within an organization. They analyse and provide advice on management methods. They conduct research to determine efficiency and effectiveness of managerial programs. For example, they might do time and motion studies to determine the quickest and easiest way to accomplish a task. They might also propose improvements in methods and systems to streamline the use of resources such as time used within an organization for internal communication.

Where do management consultants work?

Management consultants are often the key personnel in management consulting firms. They might also work for advertising agencies and other corporate sectors. They might also find employment in the public sector, in helping departments run more efficiently. Management consultants are often self-employed and sell their services by contract to both the public and private sector.

Technology facilitates multitasking in today's fast-paced world.

Families stressed over time, report says

Study shows parents work harder, earn less, spend more

By Elaine Carey and Tracy Huffman

Two working parents, rising costs, shrinking incomes and, most of all, not enough time.

It all adds up to heavy-duty stress for the average Canadian family, according to a comprehensive report released today.

The proportion of time-stressed Canadians rose by 33 percent: a fifth of all women over 15, and 16 percent of men, are stressed due to a time crunch, says the study by the Vanier Institute of the Family. The study is based on 1996 census data.

Susan Holladay, 43, a wife and working mother of three, and her husband Scott make up just one of thousands of families trying to do it all.

"It's very complicated trying to get people to the right places and make sure everyone still gets their homework done," she said. "The older ones do it on their own, but the younger one needs some direction.

"Just being there and giving each other some moral support is so important."

The study found women aged 25 to 44, holding full-time jobs and raising children, were the most likely to be severely stressed for time.

More than one-third of mothers struggle with the competing demands of jobs and families, compared to a quarter of fathers. Mothers with school-aged children were even more stressed.

No wonder. It takes the average family a year and a half—75.4 weeks on the job between two people at an average wage—to earn enough to cover the typical family's annual expenses, the report says.

Those at the bottom of the pay scale worked an average 67.5 weeks to pay for the basics, while those at the top of the scale had to be on the job for 82 weeks to pay for their premium lifestyle.

And they aren't getting ahead. Average family incomes fell by 4.4 percent or $2,561 during the '90s from a high of $58,189 to $55,628 by 1997.

Work—on the job and around the house—takes about a third of a week for the 35- to 54-year-old baby boomer cohort. Sleep takes another third.

Watching television was the next most time-consuming activity in the last third which also included eating, community work, and socializing.

And more than half of Canadians between 25 and 44 in 1998 say they do not spend enough time with their family and friends.

"Many parents at the end of the century felt they were not living up to their own ideals of family togetherness," the report says.

"They often felt guilty on the job for not spending enough time with their families; they also felt guilty about not keeping up as employees."

For the Holladays—who have three children, Carolyn, 15, Siobhan, 13, and Sean, 9—co-ordination is key.

Scott, a geophysicist who runs his own business, usually works at home. But in the last three weeks, some business travel has made scheduling a little more complicated than usual.

After school, someone needs to be home to watch Sean, said Susan, who works for a patent and trademark agency.

So if the two eldest children are busy with after-school sports, she has to arrange to have someone else there for him—or leave work early.

And those after-school activities—soccer, volleyball, badminton, track and field, and basketball—sometimes require parent drivers. Just one more stress for the Holladays.

"It's not usually a problem. But sometimes I have to rush home to drive them places," Susan said. "Co-ordinating schedules after school can be difficult."

School trips for the children are becoming more extravagant and more expensive each year. Carolyn went to Boston for a school trip and Siobhan went skiing in Vermont with her classmates.

But it is important the children have the experiences, Susan said, noting cuts to education mean they miss out on other things.

For Sean, one of the largest stresses is homework. The 9-year-old just doesn't like doing it.

When one child is dealing with a stress, often it has an impact on the whole family, Susan said.

"You have crying and so much anxiety...which really spills over into your leisure time together."

All that stress on families is taking its toll. The study shows that employees take an average of 13.2 days off to deal with family problems, costing employers at least $2.7 billion a year.

Health care costs arising from this stress amounted to another $425 million annually for extra trips to the doctor.

To cope with it, time-stressed families are buying services from others to buy time for themselves. They pay for cleaning and laundry services, buy restaurant meals and take-out food, and spend money on day care and baby-sitters.

Families with kids under 25 who buy those services spend an average $2,114 for child care, $1,122 for domestic help, $165 for laundry, and $1,181 for restaurant meals each year.

"Even so, the real work of making and sustaining a home still rests with family members, especially women," the report says.

The Toronto Star, May 20, 2000
Reprinted with permission—The Toronto Star Syndicate

Remember there are often many ways to accomplish the same task. Management is about using the resources you have to accomplish your goals. If your resources are limited, you will need to manage them carefully. You may have to make some difficult decisions so that you can do the best you can—and that might be less than ideal. Having clear priorities will make your decisions easier. Here are some strategies to consider:

Sometimes by changing your standards, you can save time. By deciding what your priorities are, you can decide which of the tasks require your best efforts, and which can be done to a lower standard. For example, if you are overwhelmed with schoolwork, spend the greatest amount of time on the assignments that are worth the most and less time on the others.

Although my purchase is not exactly what I was looking for, I decide to stop looking. The item will suit my needs and it is not worth spending more time looking for something else.

By breaking down a task into smaller segments, you might use more time, but have a better result. You can also group like tasks together. For example, when you go to the library, get the books you need for both projects you are working on, or do all the photocopying at once. Seeing a big project as a series of smaller tasks might enable you to find time. For example, you can write a paragraph or two on your project from your notes between the time you finished the dishes and your favourite TV program starts.

This project took me more time than I expected. I broke it down into manageable chunks. Now I know I did a great job!

There may be situations where you can call upon your personal resources by allowing flexibility in the fulfillment of your goal. Realize that there is usually more than one way to satisfy a goal. If you cannot find information on "the effects of competitive sports on children," change your topic immediately, not the day before the project is due.

I know I wanted to go out shopping, but I can spend time with my brothers and sister now and shop for the gift on the way home from school tomorrow.

CHECK Points

✓ Describe how technology can facilitate multitasking.

✓ Identify some resources you might substitute for time.

✓ How can breaking a task down into smaller segments help with time management?

Reflections and Connections

1. Consider the technology you have in your home. How many tasks will it enable you to do at once?
2. What rewards could you provide for yourself to keep motivated to study?
3. Evaluate your study environment. Make a list of simple improvements you could make in order to maximize your study time.

Summary

1. Time is a limited resource. Everyone has the same amount of time.
2. Time is used to accomplish the tasks of personal care, business or work, participating in leisure activities, and tending to relationships.
3. The way you spend your time is determined by your priorities. Your values and goals shape your priorities.
4. Time-management tools such as calendars, schedules, and to do lists can assist in achieving your time-management goals.
5. Develop strategies to counteract time-wasters, interruptions, procrastination, overcommittment, and unexpected events in order to use your time effectively.
6. Using note-making and listening skills, as well as managing your studying environment, will enable you to make the best use of your study time.
7. Technology in the form of electronic equipment and computers enables us to accomplish many jobs at once through multitasking.

In this chapter, you have learned to

- apply strategies for managing your time to achieve individual, family, and group goals;
- describe strategies for acquiring, increasing, and substituting your resources; and
- describe the impact of economic, social, technological, environmental, and health factors on lifestyle decisions.

Activities to Demonstrate Your Learning

1. Keep a time log for a week. Write down everything that you do and how much time it takes. Analyze the results using bar graphs. How does your use of time compare to other students?

   Compare your use of time with the statistics on how Canadians use their time. What can you conclude? Reflect on your own use of time:
   - Were you able to accomplish all of your goals?
   - Do you think you use your time effectively?
   - What strategies could you use to make better use of your time?

2. Use social science research methods to survey your classmates to determine their time-management problems and issues. Create a pie chart to show the results. Suggest strategies for supporting individuals in improving their time management.

3. Write an article for the school newspaper entitled "Good time management means good grades."

Enrichment

1. Research time-management strategies used in the business world. Describe how individuals and families could use some of the strategies.

2. Using the Internet, locate and read an article about external factors, such as overtime work, on the changing use of time in Canadian society. Share the key ideas with the class as part of the discussion of how Canadians use their time.

Managing Your Money

How often are you faced with questions and decisions about money? Can I afford it? Will I have enough money? How much money can I save? Mom, can I have some money, please! Money is a resource that can be substituted for many other resources. Money can be acquired, saved, or spent. It is often a resource that is in short supply.

By the end of this chapter, you will be able to

✺ apply strategies for managing your money to achieve individual, family, and group financial goals;

✺ describe strategies for acquiring and increasing financial resources;

✺ demonstrate an understanding of responsible money management; and

✺ explain the role of part-time employment opportunities in financial planning for adolescents.

You will also use these important terms:

budget	fixed expenses
credit	flexible expenses
employment	income
financial institutions	interest

Where Does Money Come From?

Where does your money come from? Do your parents usually provide it? Where do they get their money? Although a few people are born into families that are very wealthy, most Canadians need an outside source of income, a steady source of money to meet their needs. Families usually receive income through wages from *employment*. In most Canadian families, both parents in a family work outside the home in order to have enough income to meet the needs of the family.

As you become more independent, you will need an income to look after your own needs and wants. As an adolescent you will begin to consider how you can earn an income. You could get a part-time job and earn an hourly wage. You could start a small business and work for yourself. Offering your services as a gardener or dog-walker might be ways of operating your own small business. Once you have accumulated some money, you could also begin to receive an income from savings and investments.

There are many ways that you can earn an income.

Managing Your Money

Money is often in short supply. Unlike time, everyone does not have the same amount of money. No matter how little or how much money you have, it is important that you make your money work for you. In most cases you have used some personal resources, such as time, energy, and skills, to acquire your money. If you do not use your money in a satisfying way, then you have lost the value of two resources—the money and the personal resources you used to earn the money.

Money is the resource that is most often substituted for other resources. You can use money to buy:

Time
• Paying someone else to do a task.

Energy
• Buy a power drink or snack.

Knowledge
• Pay fees to take a course.

Technology
• Buy a computer

No matter what resource you want to buy, you need to have enough money to buy it with.

Your Attitude about Money

Just as you have unique ideas and perspectives about other areas of your life, you also have values and attitudes about money. How you feel about money is probably parallel to your opinions about other resources. You might believe that having money provides you with security and that you should save every penny that you earn. On the other hand, you may use money to meet self-esteem needs by buying expensive clothes. Someone else may feel that buying things for other people is a way of ensuring their love and friendship.

Money can be used to meet emotional as well as physical needs. If you are oriented only to the present, you will feel the need to spend money as soon as you get it. By spending your money right away, you do not allow yourself to plan and save for any future wants or needs. If you orient yourself more towards the future, you may be more inclined to save your money for the needs and wants that lie ahead.

Mellan's Money Types

The way you feel about money will have a strong impact on your relationships and on your life as a whole. Olivia Mellan, an American psychotherapist and author, believes that everyone has a specific personality type when it comes to money. Mellan states that we all lean toward one of these "money types":

Money Type	Characteristics
Hoarder	You like to save money, prioritize your goals, have a budget and review it periodically. You have a hard time spending money on yourself. You invest money for future security.
Spender	You buy yourself goods and services for your immediate pleasure and like buying gifts for others. It is difficult for you to save enough money for future financial goals. You have difficulty making budgets and sticking to them.
Monk	You view money as dirty and bad. You think having too much money will corrupt you. You identify with people who have little money. If you invest at all, you prefer socially responsible investments.
Avoider	You avoid making a budget or keeping any kind of financial record. You do not know how much money you have, owe, or spend. You feel incompetent or overwhelmed when faced with money tasks.
Amasser	You are happiest when you have large amounts of money to spend, save, or invest. You equate money with self-worth and power, so a lack of money makes you feel like a failure. You look for high-rate investments and enjoy making your own financial decisions.

What do you think?

1. What money personality type do you think you lean toward?
2. What previous experiences and values might have affected your money type?
3. Are you happy with your money personality type or would you like to alter your attitudes about money?
4. Do you think Mellans' money types describe all possible attitudes toward money?

- ✓ How can money be substituted for other resources?
- ✓ What are some common sources of income for families? for teenagers?
- ✓ Why is it important to manage your money?

Everyday Expenses

As your personal independence increases, you are more likely to make purchases to meet your personal needs and wants. As a child, your family provided you with food and clothing. Now you might have money to make your own food and clothing purchases from time to time. Most teens have some financial responsibility for their expenses. Sometimes you might have some flexibility in how you spend money.

Reflections and Connections

1. List all of the things you have bought within the last week to meet your physical needs.
2. Think about the last time you spent money on a want (as opposed to a need). What emotional need(s) were you fulfilling?
3. Do your spending patterns identify you as a present-oriented or a future-oriented person? What are the benefits of this orientation? What are some possible problems?

CASE STUDY

Ena's Story

Ena's parents recognized the fact that the family lived quite a distance from the local high school. Since her parents were unable to drive her to school, they gave Ena money for bus fare each week so that she would not be required to spend a great deal of time walking to and from school.

Although Ena appreciated the thought, she had other ideas about how she wanted to spend her time and money. She got up 30 minutes early each day in order to have enough time to walk to

school. She knew that the exercise would keep her fit and she could also save the money otherwise spent on bus fare. With the money Ena saved by not taking the bus, she was able to go to the coffee shop with her friends on Fridays after school. She was also able to save enough to buy a treat or an item of clothing every few weeks.

Ena's parents knew about her actions and decisions. They were happy that she chose to substitute her resources in this way.

What do you think?
1. Would you have made the same decision that Ena made?
2. Should Ena's parents have agreed to her decision?

Your Individual Financial Plan

No matter what your specific financial situation is, you will have some sort of income with a variety of needs and wants that you will make purchases to fulfil. The ideal situation is that you are able to pay for all of your needs and wants and have some money left over to pay for future needs. This ideal situation is rare. Most people must be very careful to ensure that their needs are met. A way to balance your needs and wants with your income is to use a budget.

A *budget* is a plan you make for spending and saving money from your income. Like most plans, a budget has a process that follows a number of steps:

Step 1: Assess your income.

Money can be acquired from many sources. You might receive an allowance from your parents, have a part-time weekend job, or do some baby-sitting from time to time. Your *income* includes any money that you take in that is available for spending. Consider the amounts that are regular and dependable when estimating your income. Money received in the form of gifts may contribute greatly to your income, but the amounts are not regular and so cannot be counted on. When estimating your income, it is better to be realistic rather than hopeful.

Step 2: Determine your expenses and spending patterns.

Ask yourself what you need to spend money on. Do you have expenses on a regular basis, or do you spend money only when a specific need arises? Teens often have regular expenses such as bus fare and lunch. These types of regular expenses are called *fixed expenses* since their frequency and amount do not often vary. Other expenses vary. Money spent on entertainment, clothing, or other optional items are referred to as *flexible expenses*. These types of expenses do not occur regularly, are not always for the same amount, and can sometimes be eliminated altogether.

A good way to determine your spending patterns and expenses is to keep track of everything you spend for a few weeks. Once you have recorded all of your expenses, divide them into the fixed and flexible categories. Determine what your fixed expenses are for a week, and then look at what you are spending on flexible expenses.

Nelson's spending

Date	Expenditure	($)
Monday	juice and ice cream at lunch	3.50
	box of fundraising almonds	5.00
	pop and pizza after school	5.50
Tuesday	lunch special	4.25
	gum	1.10
	new CD	15.95

You may be surprised at how much you actually spend. Nelson realized that he was spending close to $50 each month on CDs. He also realized that he was so busy with his activities that he did not even have time to listen to them. When Nelson realized that he was spending close to $600 a year for something he did not really use — he was stunned. He realized that if he stopped spending on CDs, he would have enough money to buy an amazing new mountain bike by next summer.

By being aware of how you spend your money, you can make informed decisions about how you would like to spend your money. You can also determine if your spending patterns reflect your values and goals. Ensuring that there is a match between your spending patterns and your values will help you to be satisfied with your personal spending habits.

Step 3: Create a spending and savings plan.

Since money is a resource that you can use to achieve your goals, you must determine your goals before you create a spending plan. Is your goal to make a major purchase such as a stereo or computer? Do you want to have a sizable amount of money in the bank so you have the freedom to take advantage of spontaneous opportunities? Is it important to you to be able to make purchases on a regular basis?

The key to a good spending plan is to create a budget that reflects your values and goals.

Begin your budget by totalling your fixed expenses for a week. Subtract the amount of your fixed expenses from your weekly income. The remaining amount of money can be used on flexible expenses. You might want to set up categories for your flexible expenses. You might also want to set a limit on spending for each of your categories.

Maryanne's Weekly Budget

		$
Income	Allowance	15.00
	Baby-sitting (after school: 8 hours @ $6.00/hour)	48.00
Total		**$ 63.00**
Expenses	**Fixed:**	
	Lunch (drinks only)	2.50
	Bus fare (to baby-sitting job)	8.00
	Flexible:	
	Savings for new stereo	20.00
	Savings for emergencies	5.00
	Entertainment	15.00
	Clothes	10.00
	Incidentals	2.50
Total		**$ 63.00**

When you are creating your budget, be realistic. There is no advantage to a plan that sets aside money for emergencies if you regularly overspend that money on snacks or treats. A realistic budget is one that you will actually be able to follow.

Although a budget is an excellent planning tool, it should only be used as a guide. Your income, needs, and goals may change from time to time. A useful budget plan is one that can be flexible in meeting your needs.

Step 4: Implement your plan.

When you have calculated a budget that you feel will meet your needs, follow it in order to reap the benefits. If you have planned to set aside a certain amount of money each week for savings, then make sure you do that. You may need to alter your schedule or time-management strategies in order to implement your budget. Perhaps you need to set time aside to actually go to the bank and deposit your savings. Continue to keep track of your spending on a regular basis.

Step 5: Assess for success.

If your plan was realistic and aligned with your values and goals, it should work for you. If you find that you are overspending in a certain area, re-examine your budget. Perhaps you did not budget enough money in a certain category. If this is the case, your plan needs to be re-adjusted.

Your budget must also be flexible enough to deal with some unforeseen circumstances. For example, if the family that Maryanne regularly baby-sits for goes on vacation for a week, she will not be able to earn that income during that week. If she has been setting aside her emergency savings as she planned, she will have enough money to get through that week. In the event that the family she baby-sits for moves to another city, Maryanne will have many decisions to make.

A budgeting plan that is successful will boost your confidence about money management. It will lay the foundation for sound money management skills throughout your life.

CHECK Points

✓ How can you balance your needs and wants with your income?
✓ What is the difference between fixed and flexible expenses?
✓ Why is saving money beneficial?

Reflections and Connections

✓ How do your spending patterns reflect your values?
✓ What reasons do you have for saving money?
✓ Identify other resources available to you to help you in implementing a budget.

In 1998, Canadian households reported spending in the following manner:

Expenditure	$
Food	5 880
Shelter	10 092
Household operation	2 362
Clothing	2 201
Transportation	6 363
Health care	1 191
Personal care	693
Recreation	2 947
Reading materials	276
Education	679
Miscellaneous	2 177
Subtotal	**36 450**
Income taxes	10 964
Insurance (including pension)	2 802
Gifts and contributions	1 144
TOTAL EXPENDITURES	**51 362**

Table produced with data from Statistics Canada

Your Family's Financial Plan

As you mature and your ability to accept responsibility increases, you will become more aware and involved with the family finances. Parents will be more likely to discuss financial issues with you. You will become aware of the family financial goals and limitations.

As an adolescent, your needs and wants are greater than they were when you were a child. You may need items that are quite expensive such as textbooks or sports equipment. After finishing high school, you may want to go to college, university, or into a specific training program. This type of post-secondary education can cost more than $10 000 per year depending on the kind of program and location you choose.

Your family has an income, but your family also has the responsibility to pay for many expenses related to supporting the family. Some expenses are fixed, others are flexible. Even within the fixed expense category, there can be some flexibility. For example, although every family needs to spend money on food, the amount of money they spend will depend on exactly what kind of food they buy and the type of store they buy it in.

In all of the categories, the values and goals of the family will determine the types of purchases they make. In the recreation category, for example, a family that values outdoor activity may spend their recreation dollars on a wilderness vacation, while another family may value physical activity and spend their recreation dollars on sports equipment.

Plan for Retirement

All Canadians who have made contributions to the Canada Pension Plan through employment are eligible for the Canada Pension. Amounts provided through this plan may not be sufficient to meet the needs of aging family members.

Some employers and employee organizations provide workers the opportunity to earn a pension. Money is put aside from every pay cheque into a fund that builds equity. Employers and employees can pay into a pension fund and be assured of a regular income after retirement.

Some workers are self-employed, or work for employers who do not provide a pension plan. The opportunity to effectively plan for a retirement income is provided by enrolling in a Registered Retirement Savings Plan or RRSP. To enable individuals and families to provide retirement income outside of a pension plan, the federal government provides this method of saving for retirement.

By participating in pension plans or RRSPs, families can plan to meet their financial needs in their retirement years.

The elements of a family budget are similar to your individual financial plan. The family has income and expenses to plan for, however, they need to plan for several individuals. Families need to plan for short-term expenses as well as for their long-term needs and goals. Some expenses are common to all family members, while others are specific to certain individuals. Long-term expenses may include savings for post-secondary education for children and separate amounts set aside for parents' retirement income.

The family financial plan includes all of the elements of a budget. An assessment of income, a determination of fixed and flexible expenses, the creation and implementation of a plan, as well as a method of assessing the plan. The financial plan of every family differs according to income, needs, and goals.

CHECK Points

✓ How are a family's values and goals related to their financial plan?
✓ What long-term expenses do families often plan for?
✓ What are some ways that families can have an income during their retirement years?

Using Financial Services

When you begin to earn an income, you will need to start using financial services. Often, income comes in the form of a cheque that can be cashed or deposited. Bills will need to be paid and cash is not always an efficient means of dealing with expenses. Banks, trust companies, credit unions, and other *financial institutions* offer a variety of financial services to meet your money management needs. In order to choose a financial institution, you need to determine what services you need and how much these services will cost.

Bank Accounts

One of the major functions of banks and other financial institutions is to provide you with a safe place to keep your money. Money given over for safekeeping is deposited into an account. The account number identifies you as

What is a financial planner?

A financial planner is a person skilled in money management.

What does a financial planner do?

A financial planner helps his or her clients to identify their financial goals and objectives. He or she does this by analysing the clients' financial records. After reporting on their analysis and discussing various options with their client, the financial planner will develop a financial plan to suit the client's needs. The plan will make recommendations concerning cash management, insurance coverage, investment planning, and retirement and estate planning. They might also help their clients to select specific financial products and investments, for example, stocks and bonds.

Where do financial planners work?

Most financial planners work for financial institutions. These institutions include banks, trust companies, investment firms, and governments. Some financial planners are self-employed and help their clients to deal with a variety of financial institutions.

the owner. Bank accounts have numerous functions. They can be used as a place to invest your money to earn interest. *Interest* is a fee paid for the use of your money. In the case of a savings account, the bank is paying you for the use of the money you have deposited with it as credit to other people. It is common for individuals and families to invest money in a savings account for an extended period of time in order to earn interest.

Another function of a bank account is to provide you with a way to pay bills and make purchases in an efficient manner. This type of working account is called a chequing account. Through this type of account you can arrange for the regular payment of your fixed expenses, write cheques, and be provided with a statement that is a record of these transactions. Money can be deposited and withdrawn frequently for daily, weekly, and monthly transactions. For example, you can prearrange to have a monthly payment automatically taken out of your account to pay a bill such as your television cable fee. This provides a convenient way of meeting your financial obligations without having to make arrangements for every bill payment. Because chequing accounts provide you with services, you will pay the bank a fee.

Financial Institutions

Bank of Canada

Established by Parliament, this government agency regulates banking throughout Canada. It is responsible for keeping paper and currency reserves for the country's financial institutions, but does not provide financial services directly to individuals.

Chartered Banks

Sometimes called full-service banks, chartered banks are non-governmental financial institutions. Services may include chequing and savings accounts, safety deposit box rental, savings bonds, loans, mortgages, and credit.

Trust Companies

Trust companies provide regular banking services as well as trustee services. A trustee manages the financial affairs of others.

Credit Unions

Credit unions are financial institutions that are owned by their members. Credit unions operate as non-profit organizations and can therefore provide financial services at a more competitive rate.

Accessing the money from your account can be done in a variety of ways:

- **You can personally go to the bank and take out some money.** This method allows you to speak with bank staff about your account and financial needs. You can ask questions and quickly verify amounts in your accounts.
- **You could write a cheque to purchase an item or pay a bill.**

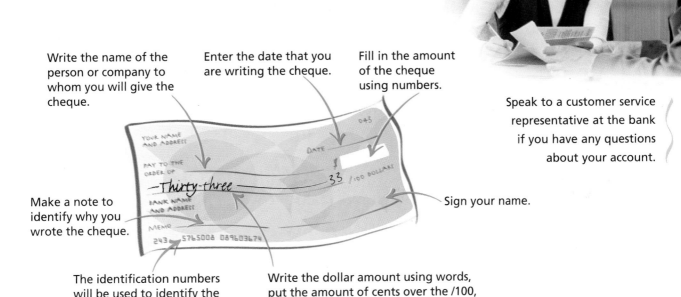

Write the name of the person or company to whom you will give the cheque.

Enter the date that you are writing the cheque.

Fill in the amount of the cheque using numbers.

Speak to a customer service representative at the bank if you have any questions about your account.

Make a note to identify why you wrote the cheque.

Sign your name.

The identification numbers will be used to identify the bank branch and account number from where the money will be withdrawn.

Write the dollar amount using words, put the amount of cents over the /100, then draw a horizontal line in the space between the dollars and cents so that no one can change the amount.

- **A debit card can be used to withdraw money from an automated teller machine (ATM).**
  A debit card also allows you to buy goods and services electronically. This type of card can be used instead of cash or cheques. Transactions are automatically deducted from your bank account. You need a personal identification number (PIN) in order to use the debit card at an ATM or at a store. Using a personal PIN provides you with security in case your card is lost or stolen.

Credit

Loaning consumers money is another service provided by financial institutions. Loans can be for small or large amounts and take many forms. Generally, financial institutions will not grant any sizable loans without collateral. When making large purchases such as houses and cars, collateral is legally held by the bank as security that the borrower will pay back the loan. If the loan is not repaid, the bank is allowed to keep the collateral.

A debit card can be used to withdraw money at an ATM or you can use it to pay for goods and services directly from your account at the point of purchase.

Smaller loans are often secured through *credit* that allows consumers to purchase goods or borrow money on the promise to repay it at some future date. A credit card gives an individual authorization for a predetermined amount of credit by a financial institution. When you use a credit card to make a purchase you are taking a loan which you agree to pay back when your credit card statement arrives. A cheque written on a line of credit that is a pre-authorized loan available for your use can also initiate this process. As with any loan, credit comes with interest costs. You must pay interest on the purchases made with your credit card or for amounts owed on your line of credit. There are pros and cons to using credit.

Pro

- do not have to carry cash
- can buy items now and pay later
- purchases are recorded
- easier than carrying and writing cheques
- might have benefits like points or air miles

Con

- credit cards can be lost or stolen
- overspending is tempting
- interest charges are high
- memberships fees may apply

CHECK ✓Points

✓ What are the differences among the different types of financial institutions?

✓ Describe the services that are provided by financial institutions.

✓ Why do consumers need collateral to borrow money?

Money is a valuable resource. It is important that you know how to use it to your best advantage. As with any resource, it can be used, managed, substituted, or saved. Ensuring that you know how best to acquire and use it will enable you to have a satisfactory financial situation.

Summary

1. Families need a dependable source of income to meet their needs. Employment is the most common means of earning an income.

2. People have spending habits that reflect their values. Money is used to meet both physical and emotional needs.

3. In order to use money effectively, a budget is a useful planning tool. Calculating income, determining fixed and flexible expenses, making and implementing a plan, and assessing that plan are the key components of a budget.

4. Families account for both short- and long-term expenses in their financial planning. Future expenses such as university tuition or retirement income require savings over a long term.

5. Financial institutions such as banks, trust companies, and credit unions provide consumers with a variety of financial services. Chequing and savings accounts are financial services that enable individuals to save money, earn interest, pay bills, and make purchases.

6. Credit, the right to use money now on the promise of future repayment, is available from financial institutions. Credit provides many attractive features for consumers. However, it does so at the cost of paying high interest on the outstanding balance.

In this chapter, you have learned to

- apply strategies for managing your money to achieve individual, family, and group financial goals;
- describe strategies for acquiring and increasing financial resources;
- demonstrate an understanding of responsible money management; and
- explain the role of part-time employment opportunities in your financial planning.

Activities to Demonstrate Your Learning

1. Using social science research methods, design and conduct a survey to determine which financial services young people use.

2. Create a personal financial plan. Assess your income and follow the five steps of the budgeting process.

3. Invite a guest speaker to talk about financial planning for post-secondary education.

4. Go to several financial institutions and investigate the services they offer. Compare services and costs. Evaluate which institution would best serve your current financial needs.

5. Talk to your friends about their feelings about money. Try to determine their money personality type.

Enrichment

1. Many organizations provide information about money management. Search the Internet for sources of money-management information and compile a directory.

2. Design a poster that promotes one piece of financial advice to appeal to people of the same money personality type as you.

Families as Consumers

You have limited resources that are valuable to you. How do you know that when you spend your time, energy, and money to make a purchase you will be getting good value? How can you tell what is really a good deal? What recourse do you have if you are not a satisfied customer?

Think about the last purchases you made. Were you happy with the product? Did you receive satisfactory service? Were all your questions answered?

By the end of this chapter, you will be able to

* demonstrate strategies for making informed and responsible consumer decisions;

* determine whether specific examples of marketing are factual or misleading;

* analyse information sources to detect marketing pressures; and

* identify the resources available for managing consumer complaints.

You will also use these important terms:

advertising	impulse buying
comparison shopping	infomercial
consumer	recycling
consumer rights	warranty
guaranteed	

You as a Consumer

The moment you spent your first penny, you became a consumer. A **consumer** is someone who purchases goods and services that have been produced by others. Individuals and families no longer produce most of what they need, so they depend on others to sell to them the goods that they need. But being a consumer can be challenging. A vast array of goods and services is available in stores, on television, and on the Internet. In today's international consumer marketplace, we import goods and services from all over the world. To make choices that best suit your needs, you need specific knowledge and skills.

Influences on Consumer Choices

As a consumer, you make decisions about how to spend your money. Every time you buy something, you make a choice among the various products available. As with all other choices in your life, there are many ways of making consumer decisions. When you make a purchase, you consider a number of factors. These factors help you think about the consequences of the various alternatives before you make your selection and hand over your money.

Income

The amount of money you have to spend is the primary factor in determining the quantity and quality of your purchases. For example, if you had a moderate income and needed new athletic shoes, you could choose to buy a moderately priced pair of cross-trainers that you could wear for all of your activities. If you had a sizable income, you could choose to purchase shoes for running, shoes for basketball, and shoes designed specifically for aerobic activities. Someone on a limited income might not buy any new athletic shoes, but would make do with old ones. Your choice would be influenced by the amount of money you had to spend.

Environment

Where you live has an impact on your consumer purchases in several ways. If you live in a cold climate you will need to buy clothing to keep yourself warm and dry. If you live in a rural area, your family might make different choices about transportation than if you lived in an urban environment. The area where you live will also have a certain number and type of shopping facilities. Suburban areas usually have large shopping malls that you have to drive to, whereas downtown areas have smaller specialty shops within walking distance. Your environment will determine both your requirements and the selection of goods and services available to you.

Personal Interests and Values

If you are really into music, you might spend much more money on CDs than someone whose life focuses mainly around athletics. If you value information and learning, you might choose to spend your money on books and software. Some people spend money on activities whereas others prefer to spend money on possessions. Your hobbies, interests, and activities will affect your buying decisions by influencing what you consider to be needs and wants.

Peer Pressure

Purchases made by your friends may also influence the types of things you buy. If all your school friends are buying a certain brand of jeans, you may also decide to follow the trend. This could be a good decision if you need to buy new jeans and you can afford the cost of this latest trend. If your jeans are in perfectly good shape and your income is limited, buying a new pair of jeans just to fit in might not be a good consumer decision. Always keep in mind that only you can determine the best use of your money.

Your personal interests will influence the types of things you choose to spend money on.

Advertising

Advertising is designed to make you want to buy a product. Advertising has one main purpose—to get you to spend your money. Few people admit to being influenced by advertising but manufacturers' research proves that when they advertise, they sell much more. Advertising may be helpful to you if it provides you with reliable information. It can also confuse you about what you want or need to buy when you are trying to make choices about your spending.

Reflections and Connections

1. In what ways does your community influence the way that you and your peers make spending decisions differently than other communities in Canada?
2. Think about your last major purchase. Identify the personal values that influenced your purchase.
3. Describe how your peers influenced a purchase that you made recently.
4. Describe how an advertisement has influenced your spending recently.

Being a Smart Shopper

Planning Purchases

Buying a product with little or no planning or ***impulse buying*** often results in dissatisfaction. In order to ensure that you are happy with your purchases, develop a plan that follows a decision-making model. Any purchase that you intend to make is based on a need or a want. The first consideration in any purchase decision is to assess whether or not your want or need is real and where it is on your list of priorities. Once you

CHECK Points

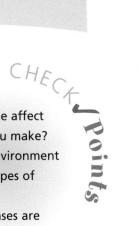

✓ How does your income affect the purchases that you make?
✓ Describe how your environment may determine the types of things that you buy.
✓ What types of purchases are most likely to be influenced by your peers?
✓ List three examples of advertising that you experience each day.

have determined your priorities, you need to look at all of the factors to consider when making the actual decision to purchase. How do you know which product will most effectively meet your needs?

All research begins with formulating the research question. You will need to ask questions to make sure the purchase will meet your standards. What are your standards for the product or service you are considering buying? For small purchases, such as a snack, the consequences are short-lived, so your dissatisfaction if you bought impulsively, and then regretted your purchase, is inexpensive. However, in most cases where you are spending a sizable amount of money on something that is more lasting, gathering information in answer to your questions will help you to feel satisfied with your purchases.

Buying Athletic Shoes—So Much to Consider

Quality
How do I judge which pair of shoes is best?

Performance
Will these shoes help me be a better athlete?

Convenience
Is this complicated lacing system really necessary?

Durability
How long will these shoes last?

Safety
Will the design of this shoe help to prevent athletic injuries?

Questions about the product you want to buy can be answered in a number of ways. Knowledgeable staff at a store can provide you with information about quality, performance, and convenience. Consumer advocate groups provide ratings about products and the manufacturers who produce them. Specialty books and magazines will often provide reliable comparisons; for example, a magazine directed at runners will frequently provide an overview of the types of new running shoes on the market and make recommendations on the quality and performance of each type. Libraries and the Internet are also good sources of consumer information.

You could also do some primary research. Survey your friends and acquaintances for their personal recommendations about products. Some people have even stopped strangers to ask them about their experiences with products. Personal experiences of individuals might be more detailed and frank than published reports. Go to the stores and examine products carefully once you know what features to look for. The more money you are planning to spend, the more important it is to spend some time on research.

Using Advertising

The millions of dollars that are spent by companies each year on advertising are factored into the price you pay for consumer goods. Make sure you get your money's worth. Although some advertising is informative, most advertising appeals to your psychological needs. Advertising strategies frequently encourage consumers to engage in impulse buying by appealing to their emotions. Resist the urge to buy on impulse and examine the appeals being used. Your psychological needs are important. Even if you are buying something you want to meet your psychological needs, you still want value for your money. You can evaluate advertising carefully to compare products and to learn what features to ask about when you speak to a salesperson. Local advertising tells you about the availability and the cost of products and services in your community.

There are many techniques and approaches that appeal to your emotions.

Advertisers, who are experts at appealing to your emotions, want you to feel that if you buy their product:

1. You will feel safe.
2. You will belong to the "in crowd."
3. You will be attractive to the opposite sex.
4. You will have good taste.
5. You will have higher status.
6. You are very intelligent.

Types of Advertising

Advertising comes in many forms. The way it reaches you is through a wide variety of media, or means of broadcasting. The three main types of advertising media are:

1. print (e.g., newspapers and magazines)
2. electronic (e.g., radio, television, or the Internet)
3. irect mail (e.g., home delivery of catalogues, coupon books, and other circulars)

What is a retail sales manager?

A retail sales manager is a person who directs the activities of a team in a retail sales store.

What do retail sales managers do?

Retail sales managers plan, organize, direct, and control the operations of businesses that sell goods and services on a retail basis. They determine what merchandise will be sold, and implement price and credit policies. Sales managers develop and implement marketing strategies including direct consumer advertising. They may be involved in placing ads in local newspapers or other types of promotions. They are responsible for ensuring the profitability of the retail operation.

Where do retail sales managers work?

Retail sales stores employ retail sales managers. They may work in stores that carry only specialty items, or they may be responsible for a specific department within a large department store. Many retail sales managers own and operate their own stores.

Before You Buy

- Deal with reputable companies, especially for mail, telephone, or electronic shopping. If you've never heard of the firm, check with the local Better Business Bureau.
- Take advantage of sales, but always compare prices to make sure an item is a bargain.
- Use unit pricing in supermarkets to compare the real cost of items in different package sizes.
- If you are being subjected to high-pressure sales tactics, walk out.
- Be sure to factor in extra charges such as delivery fees, installation charges, service costs, and postage and handling fees into your total cost.
- Ask about the seller's refund or exchange policy.
- Read the warranty to see what is covered and what you must do if there is a problem.
- Never sign a contract without reading it, if there are any blank spaces, or if there are parts that are unclear.

Consumer Tips from the publication entitled *Canadian Consumer Handbook*. Reproduced with the permission of the Minister of Public works and Government Services Canada, 2001

Comparison Shopping

Once you have decided to make a specific purchase, you will want to find the best quality at the lowest price. This will motivate you to do some ***comparison shopping*** by looking at the same product in several stores. Sometimes the same item will be priced differently according to the type of store it is sold in.

Is New Always Best?

Although we are a consuming society, there is a growing awareness of the need to reduce, reuse, and recycle. Purchasing used items has become popular and provides a lower cost option for many items. Some people think of buying used as a flexible rental because you can use an item for as long as you need it then pass it on to someone else for less money than buying new items. *Recycling* items that outlast their actual use, like formal wear or children's skates, can be a wise consumer choice.

Garage sales have become popular neighbourhood events. Families who find themselves with many things that they no longer need or use, such as items that children have outgrown or small appliances that no longer suit the family's cooking style, will have a garage or lawn sale. Often items have only been "gently used" and provide consumers with a low-priced alternative for their needs. This also allows the family who has the garage sale to recover some of the cost of the original purchases.

One person's trash is another person's treasure. Sometimes articles that are no longer of use to one person, may be exactly what another person needs.

Types of Stores

Specialty Stores

A specialty store will sell only a certain line of products, such as athletic equipment, art supplies, or electronics. There is usually a large selection of specific products, and the salespeople are usually experts who can answer your questions knowledgeably. A store that specializes in running shoes, for example, might look at how you run and ask questions about your running before recommending the type of shoe that would best meet your needs. Prices are often higher than in other types of stores.

Department Stores

A department store offers a large selection of products at a wide range of prices. This type of store is convenient for "one-stop shopping." You could buy clothes, appliances, candy, and jewellery all at one store. Department stores also provide numerous customer services such as credit and gift-wrapping.

Discount Stores

Discount stores carry products at lower prices than specialty or department stores. Selection and service may be limited and you will probably make your selections without the help of a sales clerk. Product quality may not always be consistent.

Factory Outlets

Some manufacturers sell their products directly to consumers through their own factory outlets. Prices are usually much lower than in department or specialty stores. The line of products is limited to only those produced by that specific manufacturer. These outlets may sell slightly imperfect items called "seconds."

Shopping/Warehouse Clubs

These department-oriented warehouses often charge a membership fee. Their stores provide minimal customer service and usually sell products in bulk. Prices can be lower than at conventional department stores.

Direct Sales

In direct sales someone comes to your home to demonstrate products directly to you. Traditionally, this has involved products like cosmetics and vacuum cleaners. Sales representatives bring samples of products to your home, where you can see first hand how they will look and make decisions without guesswork. Once you have chosen the products in this way, you may continue to make these purchases by phone.

Mail Order

Millions of people order merchandise from catalogues that have been delivered to their homes. This method of shopping allows you to select items when it is convenient for you and you can also place an order by telephone or by mail at your convenience. Using this type of shopping service often involves paying a shipping cost, but allows you to buy goods that might not be available in your community.

Electronic Shopping

Electronic shopping involves the use of electronic media such as television or the Internet when selecting merchandise. On some television networks, you can watch a product being advertised in an *infomercial* and then use your telephone to place an order. A service charge is usually added to the price of the product.

On-line shopping on the Internet has many benefits. Available 24 hours a day, the Internet allows you to investigate a wide range of merchandise and retail options that are not available locally or at a convenient time. For people who know what they want, the Internet is an efficient way to get it. However, only give out personal information and account numbers to companies with an established reputation.

Various retail stores sell used items. Charities might collect items such as clothing, books, and appliances, and after cleaning them, offer them for sale in their own stores. The charities do this to raise money, to provide a low-cost shopping opportunity, and to provide employment. Consignment stores, on the other hand, accept used items and sell them for the owner for a share of the selling price. Consignment stores often sell top quality items such as designer clothing for people who buy new expensive things more often but want to recover some of their purchase price. Selling out-of-print books, antique furniture, vintage clothing, or original recordings is a growing retail industry. Children frequently outgrow clothing, toys, and furniture before they show any signs of wear. Consignment shops are an economical source of items to meet the needs of families with small children.

Many people will buy interesting CDs to listen to but soon want to replace them with newer music. Selling the CDs to a second-hand store solves the storage problem, recovers some of the original purchase cost, and provides a low cost source for people who want to buy an older recording for their collection.

Reflections and Connections

1. Think about the last significant purchase that you made. What were the questions you asked about the product before you bought it?
2. Would you prefer to shop in a specialty shop, a department store, or a discount store? Why?
3. Have you every wanted to buy something that was shown on a television infomercial? Why was the product attractive to you?

CHECK Points

✓ Why are quality, durability, and performance important to consider when making a purchase?
✓ What are the advantages and disadvantages of shopping at a factory outlet?
✓ Why is electronic shopping becoming increasingly popular?

Your Rights and Responsibilities as a Consumer

Entering the marketplace as a consumer can be intimidating. As with any other type of relationship you enter into in your life, the one of buyer and seller is a give-and-take relationship. It entails both rights and responsibilities for both parties. It is important to know your **consumer rights** and what you can expect when you are interacting in the marketplace. You should also understand your responsibilities to ensure that you will not be disappointed if a purchase is not satisfactory.

Consumer Rights

- You have the right to safety with the purchase of any product or service. The Canadian Standard Association seal of approval assures you that certain groupings of products, such as electrical appliances, have been tested for safety. Numerous government regulatory bodies are in place to insure consumer safety. All products must be safe when used as they are designed to be used.

- You have the right to have complete and accurate information about any product or service that you purchase. Only with this information can you make a wise and informed decision. You have the right to choose any product or service on the market that best suits your needs. Federal laws ensure that consumer markets are competitive and offer products and services at fair market value.

- You also have the right to complain if you are not satisfied with a product or service you have purchased. You have the right to ask the seller for satisfaction.

Warranties

When purchasing most products, there are specific warranties to protect consumers. A **_warranty_** usually refers to the length of time the product is **_guaranteed_** to work, as well as to what parts are covered under the warranty. If a product does not provide a specific warranty, provincial legislation says that implied warranties apply in every sales contract. You should always check the warranty on a product before you make a purchase.

Complaining Effectively

1. First review your warranty to see exactly what is covered. Also check what the store policy is with respect to repairs and returns. Decide what you want within the limits of what is available to you. Then contact the seller to lodge your complaint and see what recourse is available to you.
2. If you still have a problem, ask for the address and telephone number of the company headquarters. Contact their Customer Service department. It may be a good idea to address your complaint in the form of a letter.
3. If the company does not provide you with satisfaction, contact the Government Office of Consumer Affairs. They will provide you with the name of the appropriate government office or consumer organization to deal with your specific problem.
4. Taking legal action should be your last resort. Small claims court is one option to consider and you might not need a lawyer. However, if you do decide to file a formal lawsuit, check with a lawyer before you proceed.

Consumer Responsibilities

There are many things that you can do to prepare and protect yourself in the marketplace. Often problems arise because you as a consumer have unrealistic expectations or are not sure of what you need or want. As you decide to make any purchase, there are a few things you should keep in mind.

- Do your homework.
  - Be sure that what you are buying is really what you want. Products purchased on impulse without any research often do not meet buyer expectations. By researching the quality and performance features of a product, you can make sound decisions.
- Act honestly and with consideration for others.
  - When shopping, do not open packages. Ask to see samples or floor models. This helps the retailer to avoid lost parts and damage to merchandise. Be careful not to soil clothing when you try it on. Do not shoplift or abuse a store's return policy.

When stores lose money due to theft, consumer prices rise for everyone. Stores combat shoplifting by installing security cameras and other electronic theft detection devices. These also add to consumer costs.

Use Products Safely

It is important to read manufacturers instructions before using a product. This may be extremely important in the case of electrical appliances, however even clothing items can be unsafe if treated improperly. It is also important to keep yourself informed of manufacturers recalls so that you can return a product to the manufacturer for necessary service.

CHECK Points

✓ Who helps to ensure that the products Canadians purchase are safe?
✓ What can you do if you are not satisfied with a product?
✓ List and describe the consequences of shoplifting.

Summary

1. A consumer is someone who purchases goods and services that have been produced by others.
2. Consumers are influenced in their choices by their income, their environment, the opinions of their peers, and by advertising in newspapers, television, and coupon books.
3. When planning a purchase, it is important to consider quality, durability, performance, and safety to determine the best price possible.
4. Personal shopping can be done at specialty stores, department stores, discount stores, factory outlets, and warehouse clubs. Consumers may also shop through direct sales, mail order, or electronically.
5. In order to make satisfactory consumer decisions, purchasers must gather information about the product, their rights, and the risks of the marketplace.
6. Used articles are becoming popular because consumers can acquire merchandise that is in good condition for a lower price than a new item while supporting conservation by re-using goods.
7. A consumer in the marketplace has the right to safe products and complete and accurate information as well as the right to complain in the event of dissatisfaction.
8. Responsible consumers are informed and knowledgeable about products, are honest and considerate, and use products safely and for the purposes for which they were produced.

In this chapter, you have learned to

- demonstrate strategies for making informed and responsible consumer decisions;
- determine whether specific examples of marketing are factual or misleading;
- analyse information sources to detect marketing pressures; and
- identify the resources available for managing consumer complaints.

Activities to Demonstrate Your Learning

1. Investigate the purchase of a product or a service by applying a decision-making model.
2. Use social science research methods to survey your peers. Find out what kind of advertising they find most appealing and how it affects their buying habits. Use the results to develop a list of tips for using advertising effectively.
3. Invite a guest speaker from a local consumer agency. Prepare questions to find out what information and services they can provide and how individuals and families can access their services.
4. Write a letter of complaint for a product or service that you were not satisfied with.
5. "Surf the net" to find on-line shopping services. Compare sites to determine how easy it is to get information about products, order merchandise, and obtain customer follow-up.

Enrichment

1. If you have a purchase to make, apply your learning from this chapter to develop a purchasing plan. Record your experiences as you follow your plan with a written, photographic, or video journal.
2. With a group, prepare a presentation for children that highlights consumer advice for spending an allowance.
3. As an individual or a class activity, write a consumer advice column for the school newspaper.

Caring for Children

Does childhood seem like another world to you, one that you have left behind? Do you remember what it was like to be a child? How do children think? Why are they amused by simple activities? How do they feel? Are you an important person in the lives of the children you know?

By the end of this chapter, you will be able to

* apply decision-making models to child-care situations;

* demonstrate the communication skills necessary for relating to children;

* describe strategies for acquiring, increasing, and substituting resources for caring for children;

* explain the various ways in which teenagers can accept responsibility for child care; and

* summarize how the needs of children change as they develop.

You will also use these important terms:

child development	role models
childproofing	safety
discipline	self-discipline
limits	socialization
parenting style	

Achieving Your Potential

When you look back at family photographs of yourself as a child, can you remember what you were like at those ages? The person you are today and the person you are becoming reflects the child you used to be. Understanding how you developed as a child can give you added insight into who you are. You can also appreciate the importance of your role when you care for children as an older sibling or cousin, as a baby-sitter, as a counsellor or leader working with children.

Childhood: The Beginning of the Journey!

Human development begins at conception. From that moment on, the transformation of cells through a unique genetic path forms the physical make-up and traits of the individual. Growth and development begin to be measured and described at birth. Every individual experiences a sequence of developmental stages from birth, through childhood and adolescence, to adulthood. Children develop physically, intellectually, emotionally, and socially as they progress through the various stages of childhood.

At each stage, children learn and master various skills and abilities that are called developmental tasks. Research has shown that there are some basic principles that influence development:

- Rates of development are individual. Some children develop more slowly, others more quickly. Each child's time frame is unique.

- Development occurs in sequence. Skills and abilities are nearly always learned in the same order. For example, most babies learn to crawl before they learn to walk.

- The areas of development—physical, intellectual, emotional, and social—are related and connected to one another. A child may be physically capable of doing something, but may not be emotionally ready to do it. For example, a child may be physically able to walk, but prefers not to, because he still needs the emotional security of being carried by his mother.

The Developmental Stages

Development appears to occur in stages. There are periods of growth and change followed by periods of stability. This means that life can be divided into stages in which growth and development are fairly predictable and, therefore, similar for all people.

Infants

From birth to the end of the first year, a baby's growth and development are more rapid than at any other stage in life.

Socially

An infant will begin to distinguish between familiar and unfamiliar faces. Once they learn this, they "make strange" when they see a face they have not seen recently.

Emotionally

Infants begin to develop trust with caregivers. Feelings of love are important in order for the baby to thrive.

Physically

Babies gain control of their senses. They focus their sight, learn to recognise voices, experience a number of tastes, and start to see the effects of their own touch. By grabbing at objects and putting them into their mouths, they begin to learn hand-eye co-ordination. They learn to hold their heads, roll their bodies, crawl, and eventually stand and walk. Great strides are made within the first year.

Intellectually

Infants begin to learn that things are permanent. Babies begin to understand that things continue to exist even when they are not in sight.

Toddlers

The toddler stage, from about one to three years, is marked by rapid changes in movement and communication.

Intellectually

Toddlers begin to acquire language and spatial concepts. The cause and effect relationship is learned. For example, "If I kick a ball, it rolls away." At this stage children begin to learn concepts such as bigger and smaller.

Emotionally

Toddlers are beginning to develop a sense of themselves. At this time they begin to have feelings of pride at being able to so something on their own. They also look for approval and support from those around them.

Socially

Toddlers are still very much in their own world. They will play beside another child, but not play together. This is called parallel play. Toddlers are starting to be independent and want to do things for themselves. Because they look to others for approval, they begin to understand that some behaviour is acceptable and some is not.

Physically

Toddlers are developing their motor skills. They learn to walk and climb. At this stage, parents and caregivers must be very vigilant as falls are very common.

Preschoolers

From ages three to five, children master the basics of movement, communication, and simple reciprocal social relationships.

Emotionally

Preschoolers are learning empathy. They are beginning to understand other people's feelings and to control their own. Verbal skills help children to talk about their feelings, especially their fears.

Physically

Preschoolers are much more in control of their bodies. They are beginning to take on more complex tasks such as riding a bicycle or using scissors to cut paper.

Socially

Children are seeking out other children for co-operative play. Preschoolers begin to make friends and may even have a best friend. They also learn how to share and to work through problems with others. They are very concerned with fairness and understand the impact of how others react to their behaviour.

Intellectually

Preschoolers are beginning to learn letters and numbers. Counting is an important skill. Verbal skills are improving and vocabulary is expanding. Children at this age have a very rigid understanding of rules and the concepts of right and wrong.

School-Age Children

In the school years from ages five to ten, children begin to apply their abilities to learning the behaviours that are specific to their society.

Physically

The school-age child continues to grow but at a much slower rate. Fine motor skills are developed to enable detailed writing and drawing. Baby teeth are replaced by permanent teeth.

Emotionally

The focus shifts from feelings of fear to concerns about success. School-age children are anxious to please others and often worry about being good academically and in other areas, for example, in team sports.

Socially

School-age children begin to choose relationships with friends. They have more activities with their peers outside the home. Children at this stage learn to take some responsibility for their own behaviour and are more sensitive and concerned with the feelings of others. This allows them to understand the reasons behind rules and guidelines, and to begin to develop a conscience.

Intellectually

School-age children are developing their reasoning and inquiry skills. They spend more time thinking about how things work and why things happen. Learning in all areas increases greatly at this stage.

The Nature versus Nurture Debate

There are many theories regarding the most important influences on *child development*. Some theorists argue that heredity plays a key role in the way an individual turns out. Other experts believe that environmental factors contribute the most to human development. Recent research on child development in the early years points to the strong relationship between early stimulation and effective development in later years.

Goals in Caring for Children

Parents and other caregivers can create environments for children that enhance and enrich their development. Each family decides what the priorities will be by their own value system. A family may value social skills highly and focus on developing social skills in their children at an early age. If their primary value is academic achievement, they may focus on stimulating intellectual development. Most families value balanced development in all areas, and will work to enhance overall child development. One of

CHECK Points

✓ Describe the highlights of child development from birth to school age.

✓ When do children begin to socialize with other children?

✓ How does early stimulation affect brain development?

Connecting the Brain

- Before birth, the brain forms billions of cells called neurons. The number of neurons and how they are arranged is basically determined by heredity.

- The brain works by making connections and communicating internally. Neurons make connections to send and receive impulses from one another. The more stimulation the brain gets, the more connections are formed. As this process develops, one neuron could link to thousands of others.

- The more connections that are made, the better the brain works. Children are able to learn language and other skills such as creativity, reasoning, and problem solving.

- Effective parenting can help make these brain connections. Research has shown that providing the child with both stimulation and security, talking to the child, playing games, and giving lots of affection, helps the child's mind to grow.

- Babies need lots of human contact and stimulation. Babies who are rarely held develop brains that are 20 to 30 percent smaller than normal for their age.

- You cannot spoil an infant by responding to his or her needs. Crying newborns who receive a quick, warm response usually learn to cry much less and sleep more at night.

- Seventy-five percent of human brain development comes after birth. To a far greater extent than other species, humans have brains that are dependent on environmental input.

- By age three, a child's brain is twice as active as an adult's. As a child reacts to stimulation, new connections are formed in the brain. The vast majority of these are formed in the first three years.

- Secure children are best prepared for school and life. When a child feels safe and secure in a loving relationship, their brain better absorbs and handles all the new input coming their way.

Parenting Tips from Invest in Kids www.investinkids.ca

the roles of parents is the *socialization* of the child. Through the guidance provided by parents, children learn the knowledge and behaviour expected by their society.

One of the primary goals for caregivers, whether they are parents, extended family members, day-care providers, or baby-sitters is to keep the child safe. Ensuring the physical *safety* of children is the most important task. *Childproofing* an environment means removing any hazards such as breakable items and using safety devices, such as electrical outlet covers to protect children. Since children do not acquire judgment until later in their school-age years, it is up to adults to ensure their safety.

Physically
When caring for toddlers, make sure all baby gates are closed to block off stairs and other potential danger areas. Make sure hot or sharp items are not left where toddlers can reach them.

Intellectually
With an infant, provide eye contact and facial movements to stimulate the brain. Smile and make sounds to engage the infant.

Socially
Support toddlers in their efforts to do things for themselves. Let them help with snack preparation, even if it means cleaning up a bigger mess later. Make sure you model acceptable behaviour.

Emotionally
Encourage and support the successes of the school-age child. Talk to them about how they are doing at school. Give praise for their efforts.

Fostering intellectual, emotional, and social development are also important in the overall care of a child. One of the key ways to influence a child's behaviour in these areas is to be a positive role model by demonstrating appropriate behaviour. All children learn by seeing and emulating the behaviour of others. For example, if a child is read to often, he will learn to like books and this will reinforce future learning.

CHECK Points

- ✓ Why is a baby-sitter responsible for a child's safety?
- ✓ How can you encourage a child's intellectual development?
- ✓ Why is it important not to lie to children?
- ✓ How can you support a child's emotional development?

Reflections and Connections

1. Describe a relationship with a child in which you think you have had an influence on the child.
2. What do you think is the most important difference between children and people your age? Why?
3. Do you miss anything about being a child? Why?

Hands On!

1. Investigate the practical aspects of child safety. Make up a checklist of things to look for and things to avoid when baby-sitting infants and toddlers.

2. In small groups, assemble a book of ideas for baby-sitters. Divide the book into four sections, infants, toddlers, preschoolers, and school-age children. For each stage of development provide a list of games and activities that will help to meet the physical, emotional, social, and intellectual needs of the child at that stage.

3. Go to the toy store and look at toys that are deemed to be suitable for infants. Identify three toys that you feel would be appropriate for this age and give reasons why.

You as a Role Model

Children form their first relationship with their primary caregivers who are usually their mother and father. Parents and other family members give the care and support that provide the foundation for lifelong family bonds. Parents meet the physical needs of children and enhance their development intellectually, emotionally, and socially.

To be an effective parent or caregiver, an individual must possess a combination of characteristics. Empathy, the ability to understand others, is an important characteristic for a caregiver to have. Understanding a child's frustration when he cannot do what he is trying to do helps a caregiver to provide support. Parents and other family members also need to be resourceful. Individuals who have knowledge about child development as well as skills in child care are more able to overcome caregiving challenges confidently. Parents and caregivers need to be realistic in their expectations of themselves and others. Caring for children is challenging. Knowing the needs and capabilities of both themselves and the children allows parents and caregivers to take the job seriously, while maintaining a sense of humour.

Communicating effectively with children is the key to developing relationships with them. Sharing activities and experiences will be easier if you can communicate at a child's level. Children also need to feel that they are being listened to, especially as they are in the process of forming their self-concept. Children are very sensitive to feelings and need to know that you mean what you are saying. Even very young children recognize when your body language does not match your words. I-messages are effective for letting a child know that you do not like their behaviour. The use of I-messages allows you to be honest while helping the child to maintain their self-esteem.

CHECK Points

✓ Identify the areas in which families must provide care for their children.

✓ Why is empathy an important characteristic skill for parents to have?

✓ How does having realistic expectations of yourself make it easier to care for children?

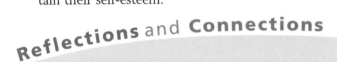

Reflections and Connections

1. Think about the customs of your family or culture. How has your family socialized you so that you fit in?
2. Describe some specific ways that you can communicate effectively with a toddler in your care.
3. Describe ways that you might use I-messages with a young child who is misbehaving.

Disciplining Young Children

As with any relationship, the interaction with children can involve problems and conflicts. It is especially difficult to manage relationships with very young children because of their lack of knowledge and inability to understand many things. In addition, personality issues can also come into play. Some children are strong-willed; others are co-operative and compliant. Depending on the match, or mismatch, with the parent or caregiver, the challenges of communication, co-operation, and relating will vary.

Children need to be protected as well as educated. Children should have a safe environment in which to explore, grow, and develop their skills. If a young child tries to do something that is potentially unsafe, then it is up to the adults to prevent an accident. The goal is to guide children so that they learn to make the right decisions and to set limits so that children are not doing things that they cannot manage yet. Identifying the limits is often easier than deciding how to guide the child's behaviour.

What do you think?

Two-year old Amy has found a pen and has started drawing on the wall. As her baby-sitter, you have many choices about how to deal with the situation. What would you do?

- **Run and grab the pen from Amy and chastise her for doing something "bad."**

- **Approach Amy and talk to her about why drawing on the wall is not acceptable behaviour and try to convince her to stop.**

- **Ask Amy to trade the pen for some crayons and give her some paper to draw on, explaining that drawing on paper is much more acceptable than drawing on the walls.**

- **Take the pen from Amy's hand and simply walk away.**

Which approach would be most beneficial for Amy's development?

Children have never been very good at listening to their elders, but they have never failed to imitate them. They must; they have no other models.

James Baldwin

Individuals develop their own approach to raising and caring for children based on the amount of knowledge they have, their values, as well as the way they themselves were raised. Some people choose to raise children in the same way as they were raised; others vow not to make the same mistakes their parents or caregivers made and choose a completely different *parenting style*. In most cases, parents and caregivers will use what worked well in the past and add new strategies as they gain experience.

Discipline is the process by which children are guided to behave in acceptable ways. Effective discipline helps children to function productively in society. The goal is to have children learn to control their own behaviour, or to have *self-discipline*. You can help guide children's behaviour in several ways:

- **Set a good example.**
  Children learn through imitation, so it is important that their *role models* are behaving in acceptable ways.

- **Praise good behaviour.**
  This is also an effective way of encouraging children. Trying to "catch" a child doing something good and letting them know that you "caught" them, gives the child the message that good behaviour gets noticed. Children who recognise the behaviour that was good, are more likely to repeat it in the future.

- **Set and enforce limits.**
  Provide children with a safe framework within which to operate. When children are very young, restrictive *limits* are often necessary for safety; for example, the child cannot leave the yard or the playground. Older children require fewer limits if they can demonstrate responsibility and self-discipline.

- **Actively correct misbehaviour.**

  Help children to understand the consequences of their behaviour. Sometimes a warning or simple explanation is all that is required to correct a child's behaviour. "Logical consequences" are an effective method of correcting misbehaviour. For example, if a child is using a toy to hit another child, take the toy away with the explanation that the toy is not meant to be used for hitting and that hitting is unacceptable behaviour.

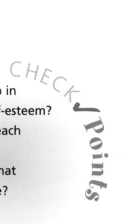

CHECK Points

✓ How can I-messages help in maintaining a child's self-esteem?
✓ Why is it important to teach children self-discipline?
✓ What are some factors that influence parenting style?

Reflections and Connections

1. When you become a parent, how might your parenting style differ from your parents' style?
2. Think of the members of your family. Who have you used as a role model?

Hands On!

1. **Think about the abilities of a toddler. Identify the limits you would set as a baby-sitter or older sibling looking after the child. Make a list of activities that you would allow the child to do and what the limits of these activities might be.**

2. **In small groups, brainstorm the kinds of unacceptable behaviours that are often exhibited by preschoolers. Keeping in mind the goals of discipline, give logical consequences for each unacceptable behaviour that you could apply in your role as a baby-sitter or older sibling.**

Families with Children

The roles of parents and children around the world are similar in many ways. Parents care for very young children in all cultures. They provide safety from harm, food, clothing, and shelter. In all cultures, parents socialize their children to be contributing members of their own society. The age at which children begin to care for some of their own needs, some of the needs of the family, and begin to take on responsibility for their own behaviour differs around the world.

In some cultures, today and throughout history, children begin to work at a very young age. Societies where hunting or agriculture is the primary source of survival require children to contribute to the welfare of the family as soon as they are able. In these cultures, children are often viewed as economic necessities within a family. The more children a family has, the greater their potential for economic gain. In industrialized societies, children often remain dependent on their families for a much longer period of time. The skills that these children will need to function in society take a longer time to develop.

When mothers are working in agriculture or in industry, they are often not able to care for their young children. Often this responsibility is taken over by older siblings. Young teenaged girls in many societies, even today, might be caring for their younger brothers and sisters much of the time. In places where girls do not go to school, or where they leave school at a young age, it is common to see young teenagers take on full-time care of infants and toddlers. When these girls marry and care for their own children, they have had a lot of experience.

In Canada today, the complex nature of our industrial urbanized society places many demands on the time of most families. Often paying someone else to care for the children is necessary. Sharing the work can free up time for parents to do other things which are necessary for the family, including having some social time for themselves. Caring for younger children is one way that you can contribute to family life, whether it is your own family or another family in the community.

Check Points

✓ Why did hunting and agricultural families need more children?

✓ Why do parents need baby-sitters in today's society?

✓ What roles do parents share throughout the world?

CHILDREN LEARN

If children live with criticism,
they learn to condemn.

If children live with hostility,
they learn to fight.

If children live with fear,
they learn to be apprehensive.

If children live with pity,
they learn to feel sorry for themselves.

If children live with ridicule,
they learn to feel shy.

If children live with jealousy,
they learn to feel envy.

If children live with shame,
they learn to feel guilty.

If children live with encouragement,
they learn confidence.

If children live with tolerance,
they learn patience.

If children live with praise,
they learn appreciation.

WHAT THEY LIVE

If children live with acceptance,
they learn to love.

If children live with approval,
they learn to like themselves.

If children live with recognition,
they learn it is good to have a goal.

If children live with sharing,
they learn generosity.

If children live with honesty,
they learn truthfulness.

If children live with fairness,
they learn justice.

If children live with kindness and
consideration, they learn respect.

If children live with security,
they learn to have faith in themselves and in those about them.

If children live with friendliness,
they learn the world is a nice place in
which to live.

Dorothy Law Nolte

Your relationship with younger children changes when you have some responsibility for their care. You might find that you become more aware of the influence that you have as a role model. You can also provide a different perspective than parents can in your conversations with children. Participating in activities you enjoyed when you were younger allows you to pass on your skills as well as the pleasure you received from the activity.

In many areas of Canada today, local governments acknowledge the important role families play in the task of raising children, and often play a role in supporting families with children.

Ontario Children's Secretariat

Secretariat Vision

The Secretariat's vision is a child-, family-, and community-centred approach to services and supports that promote the healthy growth and development of Ontario's children and youth. To fulfill this vision the Secretariat will:

- work with families, businesses, charitable and voluntary organizations, Ontario government ministries, and other governments to support parents in the healthy development of children;
- encourage early intervention in children's lives to prevent problems before they start; and
- provide families with access to information about available services and resources for their children.

Core Businesses

The Children's Secretariat has two core businesses:

- **Promoting the healthy growth and development of Ontario's children and youth**

    The Secretariat works with other government ministries and agencies to develop a unified approach to the provision of services for children. It seeks support from the private and non-profit voluntary sectors in early intervention and prevention initiatives for children, and fosters public and private partnerships to enhance services and supports for children and families.

- **Generating public awareness of services and supports available for children and youth**

    The Minister plays a key role as the government advocate, advisor, and "voice" for Ontario's children and youth. The Secretariat helps make Ontarians aware of the services and supports available for children and families. The Secretariat uses various modes of communication to help families learn about the resources available to them.

Courtesy of Ontario Children's Secretariat

Hands On!

1. **Contact your local government to find out what community services are available specifically for children.**

2. **Interview a local politician. Find out why governments are working hard at supporting families in raising their children.**

3. **Visit a parenting resource centre in your community. Find out what services they offer. Prepare a report in the form of a pamphlet that could be distributed to families in your neighbourhood.**

Practical Issues in Child Care

As with any organization that has many functions, families serve in a management capacity to get the work done. In families, parents play the role of both leaders and managers. Effective parenting requires good management skills. Decisions regarding the best ways to provide for the physical, intellectual, emotional, and social development of children are needed on a daily basis. It is necessary for parents to be aware of all the resources available to them so that they can make the most effective choices for their family.

Both human and non-human resources are necessary in order to care for children and to encourage them in their development. Many of these resources need to be acquired before you can be effective as a caregiver. Others resources are available for you to use if you know where to access them.

Knowledge and Skills

Parents and caregivers need to know what to expect from a child at each developmental stage in order to encourage proper growth and development. If a child is not progressing in a certain area, you must be able to recognize the potential problem and bring it to the attention of a doctor or other professional. The skills necessary to care for a child are numerous. Holding, feeding, and bathing an infant are critical first skills. You can learn these skills under the direction of someone who has experience. As a child ages, the skills required will change. With school-age children, you will need to be skilled at providing emotional support for school progress.

Skills Necessary for Parenting and Baby-sitting

- holding an infant
- bathing a baby
- changing diapers
- providing limits
- stimulating brain development
- teaching skills and knowledge
- preparing healthy meals

Time and Energy

Time and energy are resources in short supply in many families. Many tasks must be completed within a limited time frame. Many factors impact on the successful management of time and energy within a family. The number of children in the family will affect the amount of time it takes to care for children. Some tasks can be combined, such as bathing small children together, while other jobs, like dressing children, must be done one at a time.

Sharing tasks with others can also help save time and energy. Some tasks which may seem like a waste of time are, in fact, very important. Taking time to play with children is important in their skill development. Playing with parents and other caregivers helps children learn new skills, such as building with blocks, and provides an opportunity to enhance their confidence and self-esteem in a safe situation. You will provide a helpful role in the family by taking time to play with younger children.

As children grow they do not require as much physical care. School-age children can dress and bathe themselves and often only need to be tucked in at bedtime. They can be responsible for cleaning up after themselves and even helping with household chores. Children at this age require your time and energy in different areas. Sports and clubs are often popular with school-age children. These activities may require someone to take them to the location of the activity or to help with the organization by coaching and the like.

Play not only helps to build skills and confidence, it can also strengthen your relationship with a child.

Hours Spent On Unpaid Care of Children Per Week, 1996 Census

Population 15 years and over	Males	Females
22 628 925	11 022 455	11 606 470
No hours	7 240 690	6 696 720
Less than 5 hours	1 200 415	1 004 715
5 to 14 hours	1 196 930	1 113 775
15 to 29 hours	698 160	836 100
30 to 59 hours	385 650	788 800
60 or more hours	300 605	1 166 360

What information about the use of time does this chart provide?
Are there any indications that the roles for men and women differ when it comes to child care?

www.statcan.ca

CASE STUDY

Kersti's Story

Kersti is the mother of two small children. Michael is an active three-year-old boy and his sister Kim, who is fourteen months younger has just turned two. Kersti is a stay-at-home mom whose husband is a high-school teacher.

Kersti's day begins at 6:00 A.M. when Michael calls her. He is in the process of toilet training and wants his mom to take him to the bathroom. Just as he is finished, they hear Kim call "mama" from her room. Kersti takes the two children into the kitchen to give them breakfast. The morning meal is a challenge. Whereas Michael has a good appetite, Kim is a picky eater. The children engage in frequent squabbles throughout breakfast. Kersti tries to keep them focused on their food and plays counting games with them while they eat.

After breakfast she dresses the children while talking about all the colours in their clothing and then takes Michael to nursery school. Leaving Michael at the nursery school is difficult because he wants to come with her, while Kim wants to stay and play with the toys. Finally she is able to get Michael interested in playing with some other children in the activity centre and leaves with a screaming Kim in tow. Kersti comforts Kim in the car.

When she has calmed down, mother and daughter go to the mall to do some shopping. They need to buy groceries and diapers as well as a few clothing items. While they shop Kersti talks to Kim about what she is doing and where they are going. The shopping trip is rushed because they have to get back to the nursery school to pick Michael up at 11:30 A.M.

At pick-up time, Michael does not want to leave as he is enjoying playing with his friends. Kersti spends a few minutes talking to the teacher about the fun fair she is volunteering for this weekend before bringing the children home for lunch. She then talks to Michael about what fun they will have playing together at home and motivates him to leave the nursery school.

The children are both tired, hungry, and cranky, and Kersti finds herself frustrated with their screaming. She knows she needs to get the groceries from the car and prepare lunch. She turns on an educational program for the children to watch for a few minutes. Finally she serves up a lunch of macaroni and cheese with cooked vegetables. It is only lunchtime and Kersti is exhausted!

What do you think?

1. What resources has Kersti used in caring for her children? Could she substitute some other resources?
2. Describe how Kersti was supporting the physical, intellectual, social, and emotional development of her children.

Money

Money is a key resource in the raising of children. Children cost a great deal of money to feed, clothe, and care for. Although every family's financial resources are different, all families are responsible for providing their children with the necessities of life—nutritious food, adequate clothing, and a safe and stimulating environment.

Estimated Costs of Raising a Child to 18 Years, Canada, 1995

	$	%
Child care	52 000	33%
Shelter, furnishings, household needs	35 000	23%
Food	27 800	18%
Clothing	16 400	11%
Recreation, books, gifts, school	13 000	8%
Health care	4 600	3%
Personal care	2 600	2%
Transportation	2 600	2%
TOTAL	154 000	100%

Reprinted with permission of Home Economics Section,
Manitoba Agriculture and Food

Each family will decide how they spend money on raising their children according to their own values and goals. For example, if the family chooses to have one parent stay at home to raise the children, then the child care costs will be eliminated. Of course the family loses the income of that parent, but they may be in a financial position to do so. Families also have choices about the other categories of spending. The amount spent on food may vary greatly. Homemade baby food and meals are much more economical than commercially prepared foods. Some families may decide to purchase only organic foods for their children. This choice would significantly increase food costs. Large families could take advantage of bulk buying and save money on many items.

Clothing could also be a variable expense. Families with two or more children may save money on clothing by passing clothes down to younger children. Parents may choose to dress their children in the latest new designer fashions at a very high cost or they might purchase used clothing from a variety of sources like garage sales, consignment shops, or social service agency outlets. Some parents may decide to use their time and skills to make clothing for their children. By doing this they can have unique clothing items for their children at a very reasonable cost.

Community Resources

Families have many community resources available to help them with child care. From the time that a mother brings a newborn home, the family has access to the support of the health care system. Public health professionals are available to help new mothers cope with issues of feeding and infant care. Many community centres and religious organizations host moms groups where women can gather to share information and personal issues about child care. Community recreational organizations also provide activities such as playgroups, creative crafts, and kinder gym for small children. As children grow, various community businesses and organizations support sports activities, such as soccer and T-ball. Nutritionists and home economists are also available to help families with food and consumer issues.

Children and Consumer Protection

As children are vulnerable to many accidents and hazards, those products that are used by and for children require many safety precautions. Many national and provincial consumer agencies as well as health protection agencies provide information and guidance regarding products that may be hazardous to children.

Toys

- Toys with small parts should be labelled and kept out of the reach of children under three years of age. Children might swallow these small items and risk choking. Use the recommended age labelling on packages as a guide.
- Read all safety messages on toy labels.
- Avoid toys with sharp points or edges.

career Link

What is an early childhood educator?

An early childhood educator is a person whose profession is educating young children.

What do early childhood educators do?
Early childhood educators plan, organize, and lead activities for preschool children. They design these activities to encourage intellectual, physical, social, and emotional development. They observe children for signs of learning disabilities or emotional problems. They discuss the progress and the problems of the children in their care with both parents and other staff members. Early childhood educators are also involved in planning an appropriate environment for the children including equipment and food service.

Where do early childhood educators work?
Early childhood educators work in day-care centres and nursery schools. These can be located in community facilities such as schools and community centres. Day cares are sometimes located in places of work or office buildings. Early childhood educators may also run their own day-care centre and be self-employed.

CHECK Points

✓ Describe the knowledge and skills necessary to be an effective caregiver.

✓ How do time demands on parents and caregivers change as children get older?

✓ What control does a family have on the amount of money they require to raise a child?

Household Furnishings

• Bunk beds can seriously injure children or even cause death. Make sure that the top bunk has guard rails and that a ladder is securely attached. Do not let children play on bunk beds. Bunk beds should not be used with children under the age of six.

Clothing

• Flammable sleep wear can be a serious hazard. Sleep wear should be snug-fitting as it is less likely to catch fire if it comes near a flame. Closely supervise children around fireplaces, stoves, and candles. Choose fabrics such as polyester, which do not burn as quickly as cotton.

Reflections and Connections

1. Why do you think child care is the largest expense for families with children?

2. List some ways that you could help your family, or a family in your community to save some time or energy in the work they do in raising small children.

Hands On!

1. What resources are available to you to prepare yourself for caring for children? Check out your community resources and make a list of those agencies that could help you develop your knowledge and skills in child care.

2. In small groups, prepare a baby-sitter's tool kit. Include ideas for games and activities for children of different ages, safety considerations, special care tips, and a journal to record information for future use.

Summary

1. Physical, intellectual, social, and emotional development in all humans occurs in sequence, but at different rates.

2. The factors that influence the development of children from infants, toddlers, preschoolers, and school-age children are a combination of heredity and environment.

3. Parents and caregivers have the primary responsibility of caring for children's physical, intellectual, social, and emotional needs.

4. Effective communication and positive role modelling are the best tools for teaching self-discipline to children.

5. Although historically it was not always the case, today many governments play a role in ensuring that families receive support in raising healthy children.

6. Community resources provide many opportunities for parents to receive support in performing their role as caregivers.

7. Because children require many goods and services, families need to be aware of the many ways they can substitute and exchange resources to meet their children's needs.

In this chapter, you have learned to

- apply decision-making models to child-care situations;
- demonstrate the communication skills necessary for relating to children;
- describe strategies for acquiring, increasing, and substituting resources for caring for children;
- explain the various ways in which teenagers can accept responsibility for child care; and
- summarize how the needs of children change as they develop.

Activities to Demonstrate Your Learning

1. Investigate how research has changed our understanding of child development and brain structure. Make a list of ways that you can apply this knowledge the next time you look after a small child.

2. Compile a directory of part-time employment and volunteer work opportunities for teenagers who want to work with children in your community.

3. Contact a consumer protection agency in your community. Examine their recommendations with respect to consumer goods for children. Create a checklist of things to look for when purchasing gifts for children.

Preparing Meals for Individuals and Families

Have you ever thought about the fuel you use for your journey through life? Why do you choose the foods that you do? How can you improve your eating habits? Why is sharing food an important part of our social relationships? How does the family affect the food choices that people make? How can you share the responsibility for providing the meals for your family? Developing the basic skills for preparing meals can allow you to make a contribution to your family and share in individual and family decisions about food.

By the end of this chapter, you will be able to

✹ apply decision-making skills to planning meals to meet the needs of individuals and families;

✹ describe strategies for using food within healthy relationships;

✹ describe the variations in family meal patterns in Canada;

✹ evaluate new products and technology available for household use in preparing meals; and

✹ identify and demonstrate the practical skills necessary for preparing meals.

You will also use these important terms:

appliances	food group
Canada's Food Guide	manners
to Healthy Eating	nutrients
eating disorders	nutrition
food-borne illnesses	recipes

Achieving Your Potential

Eating is something that we all do every day of our lives. Do you know why you eat? We assume that we eat because our bodies need food for fuel, yet most of us do not give much thought to the quality of the fuel that we consume. Do you eat because you are hungry, or do you eat when it is time to eat? Are you tempted by appetizing food or food aromas to eat when you are not hungry? Understanding the reasons why you eat is the first step to forming wise eating habits.

The Fuel for Your Journey

The human body needs a steady supply of food to survive and to function properly. **Nutrients** are the chemical substances that make up food and provide the essential nourishment for the body to function, grow, repair itself, and produce energy. Nutritionists, biochemists, and physiologists study these nutrients in order to understand what each nutrient does in the body. They also determine

Food is a basic part of a balanced diet.

Fran Lebowite

Food Guide TO HEALTHY EATING FOR PEOPLE FOUR YEARS AND OVER

Enjoy a variety of foods from each group every day.

Choose lower-fat foods more often.

Grain Products
Choose whole grain and enriched products more often.

5–12
Servings Per Day

Vegetables and Fruit
Choose dark green and orange vegetables and orange fruit more often.

5–10
Servings Per Day

Milk Products
Choose lower-fat milk products more often.

2–4
Servings Per Day

Meat and Alternatives
Choose leaner meats, poultry and fish, as well as dried peas, beans and lentils more often.

2–3
Servings Per Day

Canada

how much of each nutrient your body requires for good health. Luckily, you can ensure that you obtain the nutrients you need by following the guidelines that the scientists have prepared for the Canadian government. By following the suggestions in Canada's Food Guide to Healthy Eating, you can choose foods to eat each day that will provide essential nourishment.

Canada's Food Guide to Healthy Eating organizes foods into four basic groups. The groups contain the foods that seem to belong together because of how we eat them. In fact, the foods within each ***food group*** have similar nutritional value, and are interchangeable nutritionally. For example, all foods within the Meat and Alternatives group are excellent sources of protein, a nutrient necessary for growth and repair of body tissues. You could choose to eat beef, eggs, peanut butter, pork, or dahl and get the protein your body needs. You should eat foods from all four food groups every day so that you receive all the essential nutrients.

Canada's Food Guide to Healthy Eating simplifies your food choices. For each food group, examples of servings with similar nutritional value are provided, as well as the minimum number of servings required to provide enough of the nutrients found in the foods. The Food Guide explains serving sizes for different foods. For example, in Grain Products, one slice of bread equals one serving, while one bagel, pita, or bun equals two servings. The size of servings can vary. For example, 50–100 grams of meat, poultry, or fish equals one serving. That way, a child may choose a smaller portion size than an adult to get one serving. You should choose a variety of foods within each group each day to maximize their nutritional value. Eating a large bowl containing about two servings of bran with raisins breakfast cereal, two slices of whole wheat bread in a sandwich for lunch, and two servings of pasta for dinner would give you six varied servings from the Grain Products group. It is easy to choose a variety of foods to meet the minimum number of servings from each food group every day.

Here is real food for thought: what you eat can affect how well you can think. The brain, like all parts of your body, needs adequate ***nutrition*** to function properly. Choosing enough foods from all four food groups each day can help you to perform better in your schoolwork, including remembering and solving problems. Your brain becomes dehydrated quickly, so drinking plenty of water during the day boosts your brainpower.

Above: Combine foods that you enjoy from all four food groups to make a healthy and appetizing meal.

Below: Your brain needs water. Aim for eight glasses a day!

CHECK Points

✓ How are foods grouped into the four food groups?

✓ What is the minimum number of servings a person your age requires from each of the four food groups?

✓ What are typical servings from each of the four food groups?

✓ What eating habits are recommended for better brain function?

A steady supply of food means eating regularly throughout the day. You do not have to get up during the night for meals like a newborn baby, but you should eat when you get up. Breaking that extended fast since supper the night before boosts the energy flow to your brain—eating a nutritious breakfast might result in better marks!

Are You What You Eat?

Food plays a role in your psychological well-being. Eating habits are influenced by your own body image and, in turn, affect your body image. You might feel pressure to conform to the food choices made by your friends or your family. You might make choices on the basis of personal values, such as choosing to be vegetarian, or boycotting the products of a certain country. You may have to eat food that you do not want to eat because you do not want to hurt someone's feelings. Your self-concept and your values further complicate making the choices that meet your physical needs.

Your eating habits may be influenced by your concern about maintaining a healthy body size. As you learned in the first unit, many people do not have a healthy body image because they have developed unrealistic expectations based on the ideals of male and female bodies presented in the media. Both boys and girls can develop *eating disorders* if they attempt to use overeating, purging, or not eating to achieve these ideals with their own bodies. A healthy body size is achieved by combining healthy food choices with an active lifestyle. Healthy and fit bodies of all shapes and sizes provide the maximum benefit for exciting lifestyles.

You have developed personal values related to the use of food resources. Some people have strong reactions to the flavour, aroma, or texture of foods, and want their meals to be appetizing. Do you prefer the familiar foods that are served at home, or are you influenced by advertising in the media to try new foods? Do you prefer to eat fast foods with your friends, or do you prefer to spend time cooking a meal at home? Do you eat when you are hungry or when it is mealtime? Is food a fuel that can be consumed quickly, or is food one of the pleasures of life that should be appreciated slowly? Your basic values were formed in your family, but your values can change as you experience new and different food practices, and as you take on more responsibility for your choices.

Check Points

✓ How can a teenager maintain a healthy body size?

✓ What are eating disorders?

✓ Identify some differing values that affect food choices?

✓ How can the media affect eating habits?

Eating Habits and Decisions

Do you have good eating habits? Do you think you are getting enough nutrition from your food so that you have the energy to live your life the way you want to? Poor eating habits are not easy to recognize. Many people do not realize that they could have more energy, clearer skin, brighter eyes, and even better problem-solving abilities than they do. If you compare your food intake with the recommendations of Canada's Food Guide to Healthy Eating, you can analyse the gaps in your eating habits and determine your priorities for making improvements. You have a clear gap if you are omitting a food group. You might

CASE STUDY

Gerri's Story

Gerri has found that high school has really changed her daily routine and that her eating habits have changed as a result. Because she has much farther to travel to school, she leaves home earlier than she used to, and returns home later. Friday, like most mornings, she did not take time for breakfast. She bought a doughnut and a can of cola in the school cafeteria and ate while she finished some homework and chatted with her friends about what was going to happen in their favourite TV program that night.

At lunchtime, Gerri met up with her friends for lunch as usual. Because it was cold and raining, they decided to eat lunch in the cafeteria rather than go to the fast food restaurant down the road. Gerri no longer carries a lunch to school, none of her friends do. Gerri ordered the special—a grilled cheese sandwich with fries. She chose diet cola as a beverage because she is concerned about her weight. She ate the whole lunch except for the few fries that Mary stole from her plate.

Walking home alone from school, Gerri stopped at the corner store and bought a chocolate bar for energy. She often felt tired at the end of the afternoon's classes, but she never missed a track team practice. Because there was no practice today, Gerri helped her mom to get dinner ready by peeling the potatoes. This had been her daily chore for years, but lately they have been eating pasta more often because the potatoes were not ready on time. The mashed potatoes were served with slices of mom's special microwaved meatloaf, and stir-fried frozen mixed vegetables. Gerri had a glass of milk with her dinner.

After doing her homework later that evening, Gerri sat down to watch a new episode of her favourite television program with a few oatmeal cookies and a glass of milk. As soon as the program was over, she rushed to the phone to discuss the very surprising developments with her best friend, and ate a few more cookies. As Gerri got ready to go to bed, she thought about how busy her day was, and how tired she felt. She wondered whether eating better might give her more energy.

What do you think?

1. Is Gerri eating enough food from all four food groups?
2. What improvement would you suggest Gerri make in her eating habits first? Why?
3. What factors are influencing Gerri's eating habits?
4. Apply the decision-making model to choose a strategy for Gerri to improve her eating habits.
5. What would you suggest Gerri do tomorrow to start implementing her plan?

be short a few servings in one or more groups. You may just need more variety for optimum nutrition. Setting goals for filling the gaps will allow you to develop a plan to improve your eating habits.

Deciding what action to take will be easier if you know what factors influence your decisions about what to eat. What you eat is determined by the foods that are available. You choose from the available foods that your family provides at home, or what you are able to buy in your school or community. Your personal values about food can also influence your choices, but family values can restrict your choices. Peer pressure and your self-concept might also affect your choices. Consider the various factors that influence you when you try to change how you eat. Changing the environment in which you eat might be necessary before you are able to change your behaviour.

How will you improve your eating habits? It is best to set small goals and improve one food group at a time. Perhaps you need to eat more fruits and vegetables because few are served in your home. If so, you are like many Canadians. How can you add vegetables to your daily meal plan? Compare possible strategies and select those that best meet your needs considering the many factors that influence your choices. Perhaps adding a snack of crisp, raw vegetables will enable you to eat vegetables after school, when you are away from your friends. Enjoying a fruit snack in the morning is quick and easy, and does not require the rest of your family to change if they do not want to. Develop a clear action plan based on the best strategy for improving your eating goals. Then eat better!

CHECK Points

✓ How can you determine if there are gaps in your eating habits?

✓ What should be your first priority for improving your eating habits?

✓ What are some values that can affect changing your eating habits?

✓ How can you change your eating habits if your family is not willing to change theirs?

Reflections and Connections

1. Draw a diagram to summarize how food affects your development in all areas.

2. Why might some people require more than the maximum number of servings from each food group? Why would other people reduce the number of servings down to the minimum number from each food group?

3. Compile a list of some of the causes of disagreements about food between teenagers and their families.

4. Quickly! Plan some easy breakfasts that you could eat before leaving for school.

Hands On!

1. **Make up a meal plan for a day to meet your needs by following Canada's Food Guide to Healthy Eating. In small groups, choose the most interesting breakfast to prepare and eat in class, or prepare the breakfast from your mealplan at home.**

2. **As a class, determine the major values affecting food choices for breakfast. Prepare a selection of breakfasts to meet the nutritional needs of teenagers. Invite a sample group of students in your school to taste the breakfasts and complete a survey to determine their preferences and the reasons for their choices. Compile the results and prepare a report.**

3. **In small groups, prepare a skit to promote healthy eating choices to meet the developmental needs and values of teenagers. Present the skits as street theatre, or present them to the Grade 8 class at the school in your community.**

Relating to Others

Food and Friends

What are your favourite foods? Eating foods that we enjoy is a pleasure we can all experience. Because food can be so enjoyable, some people use eating to make themselves feel better when they are sad, or lonely, or stressed. Using food this way can mask unwanted emotions but does not really meet your emotional needs. Nor does food replace friendships. Making the effort to eat with others enables people to share interests and experiences as they strengthen their relationships. Family mealtime, "doing lunch," or a romantic picnic for two are eating habits that contribute to your emotional and social development.

Eating together can be a casual social activity. Inviting someone to join you to eat lunch or for a snack are common ways of initiating a conversation. Choosing the food, and talking about it as well as the surroundings, can make conversation with someone you do not know very well much easier. For this reason, snacks are often served in situations when strangers must get to know each other and work together as a team. Organizing a picnic or a party for new friends provides an opportunity for people to spend time together and discover the shared interests that can be the foundation of a lasting friendship.

Eating together can make communication easier in a new relationship.

CHECK ✓ Points

✓ Why does eating not meet a person's emotional needs?

✓ How does food affect social relationships?

✓ How does sharing food with others facilitate conversation?

✓ Identify some traditional gifts of food.

You can communicate your affection for people by offering gifts of food. Chocolate is traditionally the food associated with love, but gifts of foods that you have prepared yourself say that you care enough to commit your time and other resources for someone. Baking cookies for the members of your group to eat while you struggle with a tough project, decorating an ice cream cake for Father's Day, or giving gifts of homemade fudge to your favourite teachers are ways of communicating with food. What message does each gift communicate? Sharing food can signal your willingness to move the relationship to a different level.

Food can be the focus of social occasions. When people get together for good times, the activities that are planned usually include sharing special foods. The custom of sharing food began when food was in short supply so that feeding others was a very generous act. Today, serving food is a social ritual that shows concern about the well-being of your guests and your pleasure at visiting with them. Pop and snacks are served when significant friends gather at your home or at a school dance. Family members might contribute their favourite dish to a potluck picnic for the family reunion. Your prom will include a formal sit-down dinner with many courses. The choice of food matches the formality of the occasion.

On special occasions, you are expected to use good **manners** to ensure that others enjoy your company. The more formal the occasion, the more formal your behaviour is expected to be. You show respect for the occasion and for other people by conforming to the standards of good manners. The basic principle of good manners is to avoid offending or inconveniencing others so that they can enjoy themselves. Knowing the rules can simplify social behaviour giving you confidence in your behaviour. You can learn what is expected by asking others before you go to the event. This is particularly useful if you are attending an event from a culture different than your own. When you

arrive, you can watch for good role models and follow their lead. If you eat your meal in the same way as others at the table you will manage fine.

Your relationships with others can influence your eating habits. Choosing to go for lunch with your friends might limit the foods that you can choose from, or require that you buy your lunch instead of eating a lunch prepared at home in order to conform to the group. If you have special food needs for religious, medical, or ethical reasons, you might find it difficult to find food you can eat without making special requests. Peer pressure can affect the choices you make about when to eat and what to eat. If your friends have different values, they might discourage you from eating the foods you want to eat, and cause conflict for you about healthy eating habits. The expectations of other people may conflict with your values and needs related to food. Disagreements that affect daily decisions related to your values can challenge your conflict resolution skills.

If you are unsure of how to eat at a formal dinner, follow the leader!

Reflections and Connections

1. What are the ways in which students in your community share food in social activities?
2. Have you been served food that you did not know how to eat? How did you cope?
3. Where do you eat your lunch? What do you eat there? Why?
4. Describe situations in which you have observed conflict related to food. How was the conflict resolved?

CHECK Points

✓ Why is food served on social occasions?
✓ Why are good table manners important on social occasions?
✓ How can peer pressure affect food choices?
✓ How can food choices cause conflict in relationships?

1. **Plan a class celebration at which foods will be served buffet style. Research the appropriate behaviour for buffets. As you enjoy the celebration with your classmates, demonstrate appropriate manners and communication skills.**

2. **Invite guests from the faculty and staff at your school for morning coffee. Demonstrate appropriate manners and communication skills to make your guests feel welcome in the classroom.**

3. **As a class, organize a potluck lunch at which you will share and discuss foods from home.**

Family Mealtime

Family mealtime is a time to spend with your family and to share your experiences with them. It is also an opportunity for your family to pass on your family culture. The foods that you eat and the rituals that are part of the meal become part of who you are. Some families have religious rituals that are part of every meal, such as saying a brief prayer. Other families have traditions that have developed within their own family, such as eating by candlelight on Sunday evening. Food taboos, the foods that some people are forbidden to eat, are learned and practised within the family. Sharing in these family mealtime rituals helps to maintain family relationships because it represents your commitment to your family.

Many people have the impression that families no longer eat together. A national study of the eating habits of Canadian families, "The Home CEO: Household Management in the '90s" should change that impression. This study found that over 90 percent of Canadians eat breakfast each morning, usually at home, and over 75 percent of Canadians eat meals together as a family in the evening. Families are more rushed in the morning. Now that both mom and dad are probably leaving home to go to work, breakfast might not be eaten together as a family. That might explain why, of all age groups, teenagers are most likely to skip breakfast. A major shift in family eating habits is that families often eat their meal together in a restaurant or fast food store, rather than at home.

Social Science Research: Morning Mathematics

The Food and Consumer Products Manufacturers of Canada and Chatelaine/Modern Woman magazines conducted a study in Canada in 1998. They asked 1600 Canadian families about eating habits. Here are the results of the questions about breakfast:

- Ninety-five percent of Canadians eat some sort of breakfast.
- Eighty-nine percent eat breakfast at home and most breakfasts (93 percent) consist of food from home.
- Forty percent of Canadians eat cereal and bread or toast each morning.
- Cold unsweetened cereal is eaten more often than presweetened types.
- One-quarter of all breakfasts include hot food that requires cooking such as hot cereal, eggs, bacon, waffles, or pancakes.
- One in ten prepare an egg for breakfast.
- One in four eat some type of fruit with breakfast.
- Two percent of Canadians eat breakfast/cereal bars for breakfast.

Questions

1. What percentage of Canadians skip breakfast?

2. Is the percentage of students in your class who did not eat breakfast today larger or smaller than the percentage of Canadians?

3. Approximately what percentage of Canadians eat a cold unsweetened cereal each morning?

4. What are some examples of this type of cereal?

5. Predict how the makers of breakfast/cereal bars would respond to this report.

The Home CEO: Household Management in the '90s study co-sponsored by the Food and Consumer Products Manufacturers of Canada and *Chatelaine/Modern Woman* magazines

Family Meal Patterns

Meal patterns of Canadians vary according to the ethnic background of the family and the amount of time spent on preparing the meal. Breakfast is usually juice or fruit, cereal with milk, toast, and tea or coffee. Some people add hot foods, such as eggs or pancakes, but these hot foods are served more often on weekends. The most common lunch foods are sandwiches, fruit and soup, but the fast food lunch of french fries, salads, hamburgers, and stir-fries are catching up. Dinner is still the major meal of the day, and consists of meat, usually ground beef or boneless chicken, potatoes, and vegetables. Some families eat rice or pasta instead of potatoes. Only four percent of Canadians are vegetarians, but meatless meals are becoming more common.

Although sandwiches are the most common choice for lunch, fast food is rapidly catching up.

Food guides help Canadians of all ethnic backgrounds choose foods to meet their nutritional needs.

Planning meals using Canada's Food Guide to Healthy Eating is easy, whatever your needs. Although the meal patterns might be different, foods from all ethnic and cultural cuisine have been organized using the four food groups and translations are available. Canada's Food Guide has been adapted by health organizations such as the Heart Association and the Canadian Diabetic Association to recommend eating plans for people with health concerns. Healthy weight loss programs use the same principles of moderate eating combined with an active lifestyle. Balanced meals can be an excellent habit for a healthy lifestyle.

The preparation of family meals is a daily responsibility. Decisions about what to eat, when to eat, how food will be prepared, and who will do the preparation reveal many of the values of the family. What are the alternatives in this decision-making process? Does your family spend money on convenience food in order to save time? Do members of your family have the skills necessary to prepare creative meals? Are the appliances available for meal preparation used regularly? In some families one person has sole responsibility for preparing the meals. In others, everyone prepares his or her own meal and eats alone. Your family is probably somewhere in the middle. You can begin to take some of the responsibility for preparing meals as you develop your skills.

Reflections and Connections

1. Describe a tradition or ritual associated with food in your family.
2. Brainstorm foods for each food group that are common in the meals of different cultural traditions.
3. Discuss reasons why people choose a vegetarian diet.
4. What role do you have in food preparation in your family? Would you like to have a larger role? Why?

CHECK Points

✓ What are the benefits of family mealtime?
✓ List reasons why families in Canada eat fewer meals together than they did in the past.
✓ Why are all Canadians able to use Canada's Food Guide to Healthy Eating?
✓ What are food taboos?

Hands On!

1. In an oral presentation, describe a food custom or tradition that is shared by members of your family. The custom might be cultural, or a tradition that is unique to your family. Explain how this custom reflects the functions of the family. Share the food custom with your classmates.

2. Choose a theme and compile a recipe booklet containing the family recipes of your classmates. Produce and sell the booklet and donate the money to an organization that supports families in your community.

3. Conduct a survey of the community to locate family-friendly eating places. Using computer software, such as Geographic Information Systems (GIS), create a directory of the map of your community. Provide the map to community organizations to share with families.

Management for Mealtime

Meal preparation requires the management of several resources to meet some very important goals. First, families need meals to meet the nutritional needs of the individuals in the family. Secondly, meals should be appetizing so that you and your family can enjoy eating as a social occasion every day. Finally, meals should be prepared carefully so that you all can enjoy the food at its best, and do not develop food-borne illnesses. These are daily goals, not long-term goals. Families use many limited resources to achieve these goals.

Food is a major expenditure for families. They must consider the family's needs and the resources that are available for meal preparation when making purchases.

The resources used for meal preparation are changing. The average Canadian family spends about $6000 a year on food, an increasing amount on ready-made food. Families have less time for meal preparation because, although 86 percent of women cook dinner each day, 60 percent of them have jobs. The "60-minute gourmet" of the past is a luxury today. Families want meals that can be on the table in half an hour! However, Canadians are becoming more interested in cooking and a wider variety of foods from around the world are more available than ever before. Also, small *appliances* simplify much of the meal preparation. The changes in resources are changing our eating habits.

Menu planning is the skill of choosing and combining foods to create appetizing meals. Choose foods that are nutritious by following Canada's Food Guide to Healthy Eating. Remember to consider the needs and preferences of the people who will be eating the meal. Age and activity level will affect the amount of food to prepare. Individual tastes, food taboos, and special diets might restrict the foods that you can include. It is better to include foods that are in season, that is, fresh foods that are currently available from local producers for the best price. Your food storage space might also determine whether you can buy fresh or frozen foods. If your family keeps staple foods, such as flour, sugar, spices, and dried or canned foods on hand, menu planning is made easier. Your creativity will combine colours, flavours, textures, shapes, and temperatures in the food for a meal that your family or friends will enjoy.

Plan balanced meals to meet the nutritional needs of family members. Nutritionally balanced meals contain foods from at least three of the four food groups. In order to achieve the guidelines of a variety of foods and lower fat choices, one idea is to think about meals one day at a time rather than one meal at a time. Plan the meat and alternate servings first. Then add bread and cereals, and fruits and vegetables to create interesting meals. Finally, add the milk and milk products as beverages, puddings, or snacks. Having a plan in mind makes it more likely that you will develop the habit of making healthy choices, even when you are away from home.

Including a variety of foods in your meals boosts the appetite appeal as well as the nutritional value.

Preparing meals that are appetizing will entice everyone to enjoy a well-balanced, healthy meal. Your appetite is determined not by hunger—we usually eat before our bodies show signs of hunger—but by your desire to eat. Variety is the key word for planning appetizing meals. You have learned that including a variety of foods from each food group is necessary for maximum nutrition. A variety of foods will also provide different flavours, aromas, and textures to add interest to the meal. Choose the colours of the foods so that the combinations are attractive. Mashed potatoes, cauliflower, and chicken breasts in a lemon sauce might smell and taste wonderful if only it looked exciting enough to try. Varying the method of preparation is effective for making meals appetizing as well.

CHECK Points

✓ How does meal preparation demonstrate resource management?
✓ List the resources to be considered in meal preparation.
✓ What is the definition of a balanced meal?
✓ Explain how to plan meals for a day.
✓ Identify the guidelines for planning appetizing meals.

Reflections and Connections

1. What is the favourite breakfast among your classmates?
2. What are some main dishes that provide alternates to meat?
3. What are some different meal patterns eaten by people in your community?
4. What did you have for breakfast today? What did you or will you have for lunch? What will you need at dinner to complete a balanced food intake for the day?

Food Preparation

The skills needed for meal preparation can be valuable resources for you, now and in the future. Once you have the basic skills for menu planning, food safety, measurement, and following recipes, you will be able to help with the preparation of the daily meals for your family. You will be able to prepare food to share with your friends. When you are older, you will be able to plan and prepare your meals independently. Using your skills for meal preparation saves money that might otherwise be spent on convenience foods, take-out, or restaurant meals, and often takes no longer than buying ready-made meals. To acquire the skills you need, learn and practise them regularly.

Food safety skills protect you from injury when you are preparing the meal and protect those who eat the meal from *food-borne illness*. Food and water that are not handled safely, following the correct procedures, can

career Link

What is a food technologist?

Food technologists conduct experiments on food and food products. They develop standards and procedures for testing products and ensure that procedures are correctly followed and data is accurately recorded.

What does a food technologist do?

A common task for a food technologist is to prepare a nutritional analysis of a food sample. The results—listing calories per portion, vitamins and minerals, fat content, etc.—will be listed on the package when the product goes to market. Food technologists also conduct scientific studies to ensure that our food is safe, our water is drinkable, and our medications are non-toxic.

Where do food technologists work?

Most food technologists work for large food production companies. They may be involved in product development as well as the quality control area. Many food technologists' jobs are also available in the public sector, working for government agencies that monitor food safety. Some food technologists work on a freelance basis, selling their services to business and industry on a project basis.

cause disease. Food safety is such an important responsibility that municipal laws regulate the safe preparation of food in public places, such as restaurants, school kitchens, and banquet halls. Mastering the principles and techniques of food safety could earn you a certificate that will be a resource that you can use when applying for a job in food service. You should follow the same basic food safety principles at home.

Recipes are used for preparing foods. Simple recipes can be memorized but the home cook or professional chef usually writes down dishes that require accurate measurements for easy reference. If you can follow instructions, measure accurately, and have a basic knowledge of cooking terms, you can use a recipe to make anything. Cookbooks are excellent sources of recipes because they often have a specific focus such as regional foods, quick and easy meals, or using your new barbecue. Magazines are a common source of recipes also. The newest way to get recipes is to download enticing recipes from the Web site of a television cooking show. Collect the recipes for your favourite dishes.

Small appliances can be used to complete some of the steps more efficiently. Buying an appliance substitutes money for the time and energy required to complete the task by hand. Some appliances, such as bread machines or pasta makers, enable you to produce foods that require so much time, energy, and skill that you would not ordinarily make them yourself. When deciding whether a small appliance would be a good purchase, consider the time and energy that will

be saved relative to the cost. Compare the cost of making the food using appliances with the cost of buying the food ready-made. Think about whether the appliance can be stored so that it is convenient enough to use regularly. Check how easy it is to clean the appliance. Carefully selected appliances can provide opportunities for you to express your creativity in new ways and have fun while you prepare meals for others.

I feel a recipe is only a theme, which an intellectual cook can play each time with a variation.

Madame Benoit

Understanding the vocabulary of cooking will enable you to follow the recipe instructions to create a new dish successfully.

Recipe Title
• often accompanied by a photograph to show you what the finished product will look like

Ingredients
• a list of the foods you will need to prepare the recipe giving exact amounts for each item

Number of Servings
• tells you how many people the recipe is intended to feed

Other Information
• alternate ingredients and/or serving suggestions, such as what other foods would go well with the recipe

Apple Breakfast Omelette
Tart to sweet apples (according to your taste) that hold shape well

This soufflé-like puff, filled with all the goodness of apples, is quick and easy enough to make on a weekday but elegant enough for a weekend champagne brunch. Make sure your guests are at the table to appreciate this impressive golden brown puff's arrival.

3 tbsp	Granulated sugar	50 mL
4 tsp	Butter	20 mL
1/4 tsp	Nutmeg	1 mL
3 cups	Thinly sliced peeled apples (about 2)	750 mL
4	Eggs, separated	4
1/2 tsp	Vanilla	2 mL
2 Tbsp	All–purpose flour	5 mL

In large ovenproof skillet, stir together 1 tbsp (15mL) of the sugar, the butter, nutmeg, and 1 tsp (5 mL) water over medium heat until melted. Add apples; cook, gently stirring occasionally, for about 5 minutes or until tender. Sprinkle with 2 tsp (10 mL) water. Remove pan from heat, shaking to loosen apples from bottom. Spread apples evenly in pan; set aside.

In large bowl, whisk together eggs yolks, 1 tbsp (15mL) of the sugar and vanilla until pale yellow. Add flour; whisk for about 30 seconds or until thickened. In separate bowl, beat egg whites with remaining sugar until stiff peaks form; gently fold into yolks in 2 additions just until blended.

Pour over apples, smoothing with spatula. Bake in 400°F (200°C) oven for about 10 minutes or until puffed and golden. Serve at once.

Makes 4 servings.

Per serving: about 210 calories, 7 g protein, 9 g fat, 25 g carbohydrate.

Tips: If you do not have an ovenproof skillet, wrap plastic or wooden handle tightly with foil.

Method
• gives step-by-step instructions of how to prepare and combine the ingredients

Cooking Temperature
• will tell you if you need to preheat the oven to a certain temperature

Nutritional Analysis
• tells about the food energy and nutrients found in the dish

Courtesy of *Canadian Living*

Recipes contain several parts to ensure that the finished product will be a success. A list of ingredients is included so that you can gather the foods you need and shop for any ingredients you do not have on hand before you start cooking. Most recipes contain both metric—the Canadian standard—and imperial measurements. These measurements are very different however, so pick which set you will use and stick with it. Measure the ingredients accurately using one set of measurements. Ingredients that retain their identity in the finished dish, such as chocolate chips in cookies, or diced carrots in a stew can be estimated. Prepare and combine the ingredients according to the steps listed. Step-by-step instructions for cutting, mixing, and cooking are provided using specific terms. Check the meaning of the terms, such as dice, fold, or sauté, and follow them precisely. Practice following recipes carefully and you will be able to prepare the exciting dishes you see in cookbooks, magazines, and on television.

CHECK Points

✓ List the types of skills needed for food preparation.

✓ Why are food preparation skills valuable resources?

✓ What are the basic principles of food safety?

✓ Why are recipes used for food preparation?

✓ What information is provided in recipes to enable you to prepare a dish?

Reflections and Connections

1. What skills do you already have for preparing meals? What skills do you want to acquire?
2. Describe the safety precautions required for preparing a recipe you will make for your family.
3. What appliances are used frequently in your family for preparing food?

Hands On!

1. Select a new appliance and evaluate its use in preparing foods that teenagers enjoy. Demonstrate the use of the appliance to your classmates.
2. Select a recipe for a food enjoyed by the members of your group. Prepare it with your group in the classroom. Identify and apply criteria to evaluate making the food instead of buying the food.

Summary

1. Healthy eating habits which meet the physical, intellectual, social, and emotional needs of individuals and families result from following Canada's Food Guide to Healthy Eating.

2. Food choices can be a means of communicating personal values and interests, showing concern and respect for others, and strengthening relationships with others.

3. Canada's Food Guide to Healthy Eating can be used to develop meal patterns that reflect the diversity of cultural traditions and the changing lifestyles of Canadians.

4. Food can be prepared more efficiently by using money to purchase food products and new small appliances designed to save time and energy.

5. Food preparation skills, including meal planning, food safety, measurement, and following a recipe are valuable resources that can be learned and practised to take on responsibility within the family and to develop independence.

In this chapter, you have learned to

- apply decision-making skills to planning meals to meet the needs of individuals and families;
- describe strategies for using food within healthy relationships;
- describe the variations in family meal patterns in Canada;
- evaluate new products and technology available for household use in preparing meals; and
- identify and demonstrate the practical skills necessary for preparing meals.

Activities to Demonstrate Your Learning

1. In groups, use social science research skills to investigate a new appliance for food preparation to evaluate whether it is a good purchase. Conduct research using manufacturer's manuals, consumer association reports, consumer surveys, and your own test of the appliance. Submit a report summarizing your conclusions.

2. Plan a balanced meal that meets the specific needs of your family using the resources available. Locate the recipes. Prepare the meal for your family, serve the meal, and enjoy it with them. Ask an adult to assess your meal planning and food preparation skills. Reflect on your achievement and suggest the next step you will take in negotiating responsibility for preparing family meals.

Clothing Care

What will you wear today? As you make your first decision of the day, what choices are you making? Are you influenced by the weather? Do you choose clothing that makes you look good? Do you have to wear a uniform? Or do you wear the latest fashions? Is your laundry stored neatly in your closet or strewn about until you are walking on it? Who does your laundry? Your clothing is a large investment of your money and in your self-image!

By the end of this chapter, you will be able to

🌼 apply decision-making skills to choosing clothing that meets your needs and goals;

🌼 plan a wardrobe that enables you to participate in the activities of your life;

🌼 identify and demonstrate the practical skills for clothing acquisition and care; and

🌼 evaluate the information, advice, and technology available as resources for making clothing decisions.

You will also use these important terms:

accessories	laundry
care label	principles of design
elements of design	wardrobe
fashion	

Dress for Success

You get dressed to go almost everywhere in life. Every morning your clothing choices are influenced by many factors. The extremes of our climate demand that you choose clothes that will keep you warm in the winter, dry in the spring and fall, and cool in the summer. The variety of activities in your life might require changes of clothing so that you will be comfortable for each activity. Perhaps you wear clothing of the same style as your friends, or you have to wear a uniform. You also want to wear clothing that looks good on you. Since you show up for school dressed, you obviously master clothing decisions every morning!

Looking Good

Clothes that are well designed look attractive and are practical. Are you developing an understanding of what styles are flattering to your body and personal colouring? Are you able to choose clothes that fit well yet allow freedom of movement for the activities you perform? Your sense of what is attractive has changed as you have matured and as *fashion* has changed. Styles that are flattering can look outdated if they are not in fashion. It can be a challenge to maintain an attractive wardrobe when fashions change. Understanding the basics of fashion design can help you make choices.

The Elements of Design

Fashion designers work with the same *elements of design* as all designers: colour, line, shape, and texture. These elements are used according to the *principles of design* such as balance, proportion, and harmony. You have probably learned about these elements in art class. Understanding how the elements and principles contribute to good design can help you to choose clothing that will look attractive on you. Analysing current fashions using knowledge of design can help you stay in fashion.

There are several ways to choose clothing colours. Choosing clothing that suits the predominantly warm or cool tones, of your personal colouring can make you look brighter and healthier. Your colour choices may be determined by the specific colour trends featured in new fashions. You might enjoy wearing the latest colours if they are flattering to your personal colouring. If you have ever been teased that you looked like a pumpkin in an orange shirt, you understand that colours have special significance because of their associations in your own life. Choose the colours that say what you want to say, about your appearance, your interest in fashion and the symbolism colours reflect.

The construction of clothes can also affect how they look on you. Some clothes conform to the shape of your body and others have a shape of their own. The seams and construction details, such as large zippers or rows

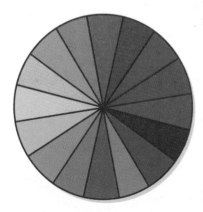

The Colour Wheel

of buttons, create lines on some garments. Lines on clothing are attention grabbers, and can make you appear wider or taller. The texture of the fabric, how it feels or seems to feel, can also affect how the garment drapes on your body. Whether you choose clothes that fit close to your body or have a shape of their own and whether you choose clingy or shiny fabrics, depends on what your parents or culture will allow, what is in fashion, and whether you want to reveal your body shape.

The Seasons Theory

The Seasons Theory of clothing selection was developed to promote the use of clothing colours suited to your personal colouring. By analysing the warmth and coolness of your skin tones and the intensity of your natural colouring, you can determine whether you are a spring, a summer, an autumn, or a winter. Palettes of carefully selected colours are recommended which will be the most flattering for each season's personal colouring. Perhaps you have "had your colours done" and can describe your season. If so, you will have a copy of the colours in your seasonal palette to use when shopping.

The Seasons Theory has been the subject of several books which you and your friends can use to analyse your own colouring. Some designers use the Seasons Theory when designing their fashion lines, and many clothing stores sort their clothing into palettes. Some salons and clothing consultants are trained in colour analysis so that they can help you choose the clothing colours, jewellery tones, and make-up that is best for your own colouring.

Reflections and Connections

1. Describe your favourite colours and colour combinations.
2. Identify some common colour associations in your community.
3. Survey your class to determine the most common clothing shapes and styles.
4. Summarize the lines, shapes, and textures that are fashionable in school clothes.

Team colours enable everyone to know who's on our side!

Using the Principles of Design

The principles of design are the guidelines that designers use when working with the elements of design. The principle of proportion describes the relative sizes of parts of the outfit and of the parts of the body. Consider the relative sizes of the upper body and the lower body when you wear your pants at your waist instead of at your hips. Changing the fit also affects the proportions of an outfit. An outfit that follows the principle of balance is attractive to look at because no one part dominates although parts may be different. The principle of harmony means that parts of an outfit should create a consistent image. Wearing running shoes with a dressy outfit would not be harmonious. Ignoring the principles of design challenges our thinking about what is attractive, but can result in unusual fashion trends.

By analysing the latest fashion ideas using the elements and principles of design you can determine which aspects you can use to enhance your appearance. Ideas about what is attractive change constantly. You may find that your body image is influenced as much by fashion as it is by your physical development. You probably will not replace your clothes every year. Knowing the specific details which are important for this year's "look" will enable you to adapt your wardrobe to keep up-to-date within your budget and to suit your lifestyle.

Which shorts fit best? Even team uniforms styles change over time.

CHECK ✓ Points

✓ What are some examples of changing proportion in dress?
✓ How can you choose clothes that are in proportion for you?
✓ How can you use harmony to put together a good-looking outfit?

Reflections and Connections

1. What are your favourite styles? What do you think your choices say about who you are?
2. What associations do you have concerning clothing design; for example are there colour associations, or associations concerning length of skirts or pants?
3. What are some other ways of using clothing design in your community? What does that clothing say about the person?

Enhancing Your Performance

The fact that Canadians have active lives in the extremes of our climate can be credited to smart clothing choices. Our long, cold winters require that we dress to protect ourselves from the cold when we go outside. Dressing in several layers of clothing ensures that you will be warm outside, but able to remove clothing so that you are comfortable in the warmth of your home or school. An absorbent inner layer, such as underwear or a T-shirt, wicks the moisture away from your body and creates a layer of heated air next to your body. Adding a spongy layer, such as a sweater or fleece top, holds pockets of air for more insulation. A breathable but closely woven outer layer protects against the wind and keeps moisture out. The fact that extra warm blood flows to your brain and your hands means that adding a hat and gloves is necessary to avoid heat loss from your extremities.

Before you know it, cold winters give way to hot summers. Keep cool by wearing loosely fitted clothing made from open weave fabrics. Natural fibres are more absorbent and will wick the perspiration away from your skin so that it can evaporate. Concern about UV radiation has changed our style of dress. You reduce your exposure to the sun by wearing clothes that cover your arms, shoulders, and legs, but allow for air circulation. This is cheaper and more effective than sunscreen lotion. A brimmed hat will protect your face and neck, keep your head cool, and help shade your eyes from the sun.

In many parts of Canada, you need protection from the insects during the warm months. Wearing long sleeves and pants allows you to enjoy the outdoors by protecting you from mosquitoes in the evenings. During the day, avoid dark colours that attract bugs, including blue jeans, unfortunately. If you are hiking or working in the bush, tucking your pants into your socks adds additional protection from ticks.

Staying dry is a concern in many parts of the country in the spring and the fall, especially if you must walk to get to school. It rains,

Are these shoes made for walking? Some fashionable clothing is not designed for comfort or practicality.

✓ How does dressing comfortably for the environment enhance your performance?

✓ Explain how to dress for extremely cold weather.

✓ How should you dress if you have to be outside in the hot sun?

✓ How do the designs of footwear vary for different activities?

Fashion condemns us to many follies; the greatest is to make oneself its slave.

Napoleon Bonaparte

on average, 100 or more days a year. Investing in an all-purpose, water-repellent jacket and shoes will make your walks more comfortable in all weather. Waterproof clothes and footwear are also available for those of you who live in wetter areas or must work outside in the rain. Do not forget a hat or a hood to protect your hair!

Clothing can be designed to meet the specific requirements of an activity. Your choice of bathing suit depends on whether you want to look good on the beach with your friends, or compete on the swim team. Classical and modern dance requires both boys and girls to wear clothing designed to maintain a smooth body line. For sleeping and for physical education class, you need comfortable, absorbent, and non-restrictive clothing, but you probably have different clothing for each activity. Clothing designed for specific sports have become everyday fashions for teenagers in many parts of Canada.

Footwear provides protection and support for your feet. Water-repellent finishes and insulation provide protection from the weather. Safety is determined primarily by the design of the sole of the footwear. Technological developments mean that you can buy special footwear designed to allow safe movement in almost any activity. You may value style more than safety for walking to school, but winter ice demands traction. Choosing footwear suited to the activity can reduce injuries and, perhaps, enhance your performance.

Reflections and Connections

1. What is more important to you—comfort or style?
2. What clothing should you consider adding to meet the safety and comfort needs of your environment?

Are You What You Wear?

Clothing choices reflect your personality and your mood. Your clothing choices project your image of who you are and what is important to you. Have you ever stopped wearing an article of clothing because it did not suit who you are anymore? Do you dress differently than your classmates? Do you find that some clothes make you feel better than others? Some students dress differently for cultural reasons. Other students develop a personal style based on their desire to be different. Your personal style, like your identity, develops over years of practice. Changing how you dress can affect how others see you and influence your relationships with others.

Clothing Won't Tell You If Teen Is Violent

By Maryln Schwartz

What you wear to school has always been an eternal struggle between parents and kids. Kids are going to rebel. Parents are going to yell. But with stories of a "Trenchcoat Mafia" surrounding the horror of two boys, who killed students and bombed their high school, suddenly the question of school dress is once again in the spotlight.

People keep wanting to know why someone didn't realize something was wrong when a gang of students appeared every day in black trenchcoats and berets and wore them before, during and after classes, even in hot weather. Where was the principal? Where were the teachers? What were parents thinking? Why didn't they notice those black trenchcoats?

I can understand the concern. I don't fully understand the question. Have these people been looking at what some kids are wearing to school these days? I live near a high school and see kids coming and going all the time with clothes and hairdos that would have gotten them committed to an institution when I was in high school. Today, it's just the way things are.

Girls spike their hair on one side and shave it on the other. Teenagers have tattoos and pierce almost every part of their bodies and wear rings all through those piercings.

Since television reporters were talking non-stop about those trenchcoats, I made it a point to watch the kids arriving at my neighbourhood high school Friday morning. If you asked me to pick out the ones most likely to be homicidal based on what they were wearing, I would have no way of knowing. Many looked just like average kids. Quite a few looked bizarre. Boys wore makeup and strange hairdos; girls were in military-looking outfits with short, short buzz haircuts. And then there were the girls with fluffy hairdos and designer handbags.

School is not what anyone over 35 remembers.

I am not trying to make light of or take away from the terror of what happened in Colorado, but we can't take an easy answer to how this would have been prevented by watching for signs of rebellion in teenage dress.

On the same network where parents were blaming the principal for not paying attention to the trenchcoats, Diane Sawyer was interviewing Dennis Rodman. If you see him with all of his body piercings, strange hair, tattoos and cross-dressing being treated as if he looks just a little eccentric, who is going to think some rebellious teenager looks different? Dennis says he is misunderstood. So do a lot of teenagers. So what is new?

So what is the answer? Will school uniforms be a solution? I asked some of the kids going to school what they thought of uniforms. One young man who was dressed as if he could skip fifth period and go straight to fighting in Kosovo said no.

"You think I look weird, but all those bowheads (girls who wear bows and dress like their mothers want them to) and their alligator boyfriends (boys in alligator shirts) look awful to me. I don't want to look like them in any way." This was a very polite kid. He said he wanted to go to college and study computers. I asked if he thought his offbeat look would hurt his later career. He said no. Being weird only helped Dennis Rodman.

The Toronto Star, 05-06-1999

Hands On!

1. **Create a collage of the styles that reflect who you are. Add words that indicate what you believe your clothing says about you.**

2. **Organize a class debate on the resolution: Be it resolved that you cannot judge people by their clothing.**

3. **Analyse current fashions to compare fashionable colours, the shapes, lines, and textures that are in style for school wear and for dressy occasions. Describe the current rules for using proportion, balance, and harmony. Find examples of outfits that are fashionable and create a display of garments that are available to enable someone to dress fashionably at a reasonable cost.**

Dressing the Part

Choosing what to wear is an important part of planning for every occasion in your life, even going to school each day. You sense that what you wear might have some impact on how well your day unfolds. Some situations have dress codes, but for most occasions you have to decide for yourself what will be the best outfit to meet your needs and the expectations of others.

Among teenagers, styles of dress often identify the subculture that you belong to. Groups, such as the "jocks," dress in a recognizable style that reflects their common interests. Dressing in similar styles can be a way of conforming to a group in order to belong. Wearing similar styles could mean that friends share the same values. The reason for conformity in dress can also be as simple as shopping in the same stores where the clothes are similar in style.

Wearing clothing that reflects your individuality could mean that you stand out as different. People assume that someone whose clothing is unusual also behaves in unusual ways. Is this true? School uniforms are based on the belief that conforming to a conservative style of dress might result in more conservative, conforming behaviour in schools. In Canada, uniforms are associated with elite schools because they were once worn only at private schools. There is no social science evidence that uniforms change the behaviour of students, but some school districts choose uniforms for students because of the clean-cut image and the ease of identifying students.

You might choose an unusual style in a desire to be different or to conform to an image that suits your self-concept.

Your clothing should be appropriate for the occasion. A dress code describes appropriate dress for school or for work. Tradition requires that you wear very dressy clothes if you go to the formal because it is such a special event. It is a custom at some schools for boys on school teams to wear a tie on game day. Common sense says that running shoes and sweats are appropriate for your physical education class. Dressing appropriately tells others that you are there to participate in the activity that has been organized. Wearing clothing that is suitable for the occasion shows your respect for other people who will be attending.

Does conforming by wearing a school uniform mean that students are better behaved?

School Uniforms Promote Spirit, Equality, and Pride

By Kimberly Ahing

Regarding uniforms, we, the students' council at Regina Mundi Secondary School, feel it is important to address the issue from a student's perspective.

Uniforms are a community builder—no question. In high schools, there are obvious splits among different groups. Tension between these groups sometimes prevents friendliness between them. Uniforms are the first step in crossing these borders between people. Students will no longer be prone to judge each other based on what brand of clothing they have. Eliminating this prejudice can only help encourage a greater sense of community and acceptance.

School sports teams demonstrate the positive value of uniforms. Members of the team are required to wear the team uniform. The uniform gives the team a feeling of pride. The relationships among teammates can become some of the closest of the high school years. The team uniform is one part in helping develop this spirit and pride.

Many students are required to wear uniforms at their part-time jobs. Getting jobs shows we, as adolescents, are maturing. We wear the uniforms required by work with pride and dignity. As teenagers, our job is going to school daily. If uniforms are required by school policy, then it is our job to wear them.

The issue of safety is still an important issue at any school. Schools are not immune to strangers, who can easily gain entrance and pose a threat to safety. Having uniforms will eliminate this problem, because trespassers can be more easily identified.

Individuality is a key issue for many students, but students, as they mature, begin to realize they do not need to dress differently to be individuals. Uniforms do not change a student's personality and mindset. They do not change their creativity, their heart or their soul. They do help students experience wider acceptance within the school community.

Uniforms are meant to welcome students and visitors, not set them apart. It is our intent to build the school community to its highest standards and uniforms are a big step toward this goal.

The London Free Press, 01-29-2000

You decide what you will wear. When you make your decision each morning you are deciding what statement you will make about yourself, about your willingness to participate in the activities of your day, and about your relationship with the people you will meet. The challenge is to balance conformity and self-expression so that you retain your individuality as a member of a group.

Dressing up for the formal makes the evening a special one for teenagers.

CHECK Points

✓ How does clothing express your personality?

✓ Why do some teenagers choose such a different style of dress?

✓ What are some reasons why people conform to the styles worn by their peer group?

✓ Why is it good manners to dress appropriately for any occasion?

Reflections and Connections

1. List the occasions in your life when specific clothes are required. Compare lists with your classmates.

2. Identify the pros and cons of wearing school uniforms.

3. Compile a list of characteristics of appropriate and inappropriate dress in your community.

4. Describe the clothing style of a subculture in your community.

Hands On!

1. Conduct an experiment. Wear an outfit to school that does not reflect your usual style of dress. Perhaps it is the style of another culture or subculture, or too dressy for school. Wear the outfit without comment and observe the reactions of others. Form conclusions about the effect of clothing on your relationships.

2. Investigate the appropriate clothing for a specific social, cultural, or work situation in your community. Write a brief, point-form dress code and attach a picture of someone dressed appropriately.

Your Clothes and Your Family

Your clothing tastes began to develop in your family. You have probably inherited your personal colouring from your parents and learned from them the colours and colour schemes that are flattering. Perhaps you enjoy wearing red, and choose crisp shirts instead of soft T-shirts because you remember your favourites amongst the colours, textures, and fit of the garments that your parents chose. You might have learned to choose clothes that are easy to care for because your family does not have time for hand washing or ironing. Your family may also determine the types of clothing that you are permitted to wear because of your culture. You have learned your values concerning dress from your family.

You have also been influenced by your peers and by the media to like certain styles of dress. Fashion is an important part of the popular culture, and is often related to tastes in music and leisure activities. Fashion creates excitement because it is ever changing. The desire to have new clothes might reflect the belief that you too are changing. There might be conflict with parents because of your different values related to clothes. For example, parents and daughters might disagree on what is modest. Understanding why you want to dress the way you do can help you to make choices more easily and to negotiate with your parents to resolve any conflict.

Your family resources will determine to some degree the amount of freedom you have to wear the current fashions. Being a fashion leader who is the first to adopt the latest styles can be very expensive. Planning your *wardrobe* once or twice a year can enable you to dress appropriately for the various activities in your life and look up-to-date with fewer clothes. If members of your family sew or knit, you can acquire the skills to design and make original clothes for yourself. You can also learn these skills in school. Finding creative ways to expand your resources often results in a unique look.

CHECK Points

- ✓ How do families influence an individual's clothing tastes?
- ✓ How can clothing be the cause of conflict in families?
- ✓ What are some ways of updating your wardrobe for less money?

Reflections and Connections

1. Describe the values that you have learned in your family concerning clothing. Do your friends share these values?
2. What are some examples of conflict with your family over clothing? How were the conflicts resolved?

Clothing Care

Accepting responsibility for the care and maintenance of your own clothing pays off in several ways. Taking care of your own clothes contributes to the household work. You have learned that doing chores reflects your increasing independence within your family. Clothes that are clean and pressed look better so that you look more attractive, and you will have more choices of clothes each day if they are all ready to wear. Finally, clothes that are well maintained last longer and can save you money on replacements. Caring for your clothes shows that you are responsible for yourself on a daily basis.

Clothing care begins when you plan suitable storage for your clothes. You need space to hang up your shirts and your dressy clothes. Some people like to hang pants on hangers and others prefer to fold them. Knitted garments such as sweaters, sweatshirts, and T-shirts last longer if they are folded and stored in drawers or on shelves. Smaller items like socks and underwear usually go in drawers or baskets. You can be creative in finding storage space for your clothes, but keep it simple. Easy access to the drawers, shelves, and hangers increases the chance of you actually putting your clothes away.

Clothes keep their shape and last longer if they are stored properly. That means every day! Clothes that will be worn again, such as sweaters or pants, should be shaken out and hung up or folded when you take them off. Clothes that need washing can be put into a hamper ready for laundry day. There does not seem to be any research on the effects of walking on your clothes for a week, but the results are probably not good.

You can decide to do your **laundry** on your own or to take on the responsibility for the family laundry as well. Learning laundry skills allows you to take special care of your clothes. Talk to your family about a good time for you to do the laundry. You will need to do several loads, so allow a few hours. If your family uses a Laundromat, you may have to co-ordinate with others for the trip. If your family has an energy saving plan, it is cheaper to do laundry in off-peak hours, usually the weekend, (or the middle of the night). As you know, forming regular routines results in efficient habits, so you might want to stick with your personal laundry day.

Begin by sorting your clothes according to care required. There will be a **care label** with the universal symbols to tell you whether a garment should be dry-cleaned, hand washed, or, easiest of all, machine-washed. Sort washable clothing into two loads: white and light-coloured items, and darker-coloured items. To avoid "fallout" on all of your clothes, check all your pockets for tissues, money, homework, notes, or bus passes. Removing stains should be done as soon as you notice the stain, but pre-treating washable stains can be done when you are sorting.

Extend the life of your favourite clothes by sorting them according to colour and care required.

Follow the Signs

These symbols tell you which procedures to use or avoid when washing, drying, and dry cleaning.

	Stop	Be careful		Go ahead
Washing	Do not wash	Hand wash in cool water	Machine wash in cool water at a gentle setting – reduced agitation (30°C)	Machine wash in warm water at a normal setting (50°C)
		Machine wash in lukewarm water at a gentle setting – reduced agitation (40°C)	Machine wash in warm water at a gentle setting – reduced agitation (50°C)	Machine wash in hot water at a normal setting (70°C)
Bleaching	Do not use chlorine bleach	Use chlorine bleach with care		
Drying		Dry flat	Tumble dry at low temperature	Tumble dry at medium to high temperature; Hang to dry; Drip dry
Ironing	Do not iron	Iron at low setting (110°C)	Iron at low setting (150°C)	Iron at high setting (200°C)
Dry Cleaning	Do not dry clean	Dry clean – with caution		Dry clean

Only iron *clean* clothes to avoid setting stains or grime

Follow the directions on the washing machine for machine washing. Measure detergent accurately into the washer according to the instructions on the package. Adding fabric softener reduces static electricity in your clothes so that clothes do not cling and you get fewer static shocks in the winter. Place the clothes in the washer without overloading. Set the water temperature on Warm for most clothes, adjust the water level for the size of the load, and turn on the washer. Did you know that one wash load takes about as long as one sitcom?

Clothes can be dried in a dryer or on the clothesline. If you remove the clothes from the dryer as soon as they are dry, clothes will be less wrinkled and you will be conserving energy. Clothes dried on a clothesline smell fresh, but are more likely to require ironing. Hanging up the clothes or folding them and putting them away as soon as they are dry will also reduce wrinkling.

Ironing removes wrinkles from your clothes and smoothes the fabric. Many fabrics today do not wrinkle if cared for properly, and some clothes are designed for a crumpled look so you might not have much ironing to do. Check the care label and sort the clothes according to the correct iron temperatures. Use steam settings for higher temperatures or a spray of water for lower temperatures to provide the moisture that makes ironing easier. Iron the small details like collars, cuffs, and yokes first, then finish with the sleeves, legs, or the body of the garments. Hang up your freshly ironed clothes immediately.

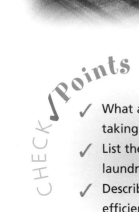

CHECK Points

- ✓ What are the advantages of taking care of your clothes?
- ✓ List the steps in doing your laundry.
- ✓ Describe how to iron a shirt efficiently.

Reflections and Connections

1. How can you save energy when doing your laundry?
2. Brainstorm ways of expanding clothing storage space.
3. How can you reduce the amount of ironing your clothes require?

Hands On!

1. Investigate the clothing storage needs of a typical teen in your community. Design inexpensive ways of organizing clothing storage to meet those needs using the resources available.

2. Determine the most common types of stains and investigate ways of removing them from fabrics used in student clothing. Publish a one-page flyer to inform students and their families.

3. Investigate clothing care products to compare the cost per use, the convenience, and the impact on the environment. Compare with a homemade recipe or traditional method. Present your conclusions in a poster using data from your investigation.

Managing Your Wardrobe

A well-planned wardrobe means that you will never have to say, "I have nothing to wear!" as you turn down an invitation. You will have the clothes that you need for the different activities in your life, summer, fall, winter, and spring. You will be appropriately dressed and ready to enjoy your life. You can look your best and be fashionable using the resources available to you.

Managing your wardrobe, like all management tasks, begins by identifying your goals. What are your clothing needs? List the activities that you must dress for and describe the criteria for choosing clothes for each activity. Remember that you will need clothing for all seasons, but some items can be worn year round. You will also identify the colours and styles that are flattering to you as well as those that are fashionable. Consider how important it is to you to dress according to the fashion in your community. Your wardrobe goals should be based on your practical, decorative, self-expressive, and social needs.

Planning Your Wardrobe

Identifying your needs is the first step in planning your wardrobe. An inventory of all of your clothing might prove that you already have many of the clothes required to meet your needs. Now is the time to recycle clothing that does not fit your body, or your personality. Some clothes can be remodeled into more usable clothing. Analyse the clothes to determine possible colour schemes and new ways of combining clothes into a variety of outfits. Do not forget to check your socks and underwear. A gap analysis is the final step where you identify the items that you need to add to complete your wardrobe.

CASE STUDY

Kate and Rhiannon's Story

"Are you coming to the Athletic Banquet on Friday, Rhiannon?" Kate asks as she catches up to her friend on the way home from school.

"I haven't decided yet," answered Rhiannon. "I'm not an athlete, and I don't really have anything I can wear."

"First of all, you don't have to be an athlete to attend—just to win best female athlete and Melanie's already got that covered! You've got to come just to see Andrew dressed up and in a tie. And you know he'll give you a flower," Kate kidded her friend. " Secondly, I can't believe that you have nothing to wear. You have more clothes than anyone I know. What about the navy blue dress you wore to that family party you went to?"

"I can't wear that. It's dirty and I don't think it can be washed. I don't have time to get it cleaned. What are you going to wear, Kate?"

" I am trying to decide between the green silk dress I made for my cousin's wedding or the red dress with the beads that I wore to the semi-formal. Or I might borrow my sister's blue dress; I think it fits me."

" I think you should wear the green dress or the blue dress. You don't want to wear the same dress to a school thing twice."

Kate shrugged her shoulders. "Why not, if it looks good. I can't afford to buy another dress when I have two or three already. I need my money for a new racing bathing suit. They are really expensive for such a small piece of cloth! Why don't we have a look at your dress and see if it can be washed? That's if we can find it in your room! If not, maybe you can borrow the dress I'm not wearing."

What do you think?

1. Why doesn't Rhiannon have anything to wear to the banquet?
2. What resources does Kate have that enable her to have a suitable dress?
3. Is it acceptable to wear the same special-occasion outfit twice?
4. Role-play the conversation between Keven and Andrew about what they could wear to a semi-formal banquet. Do boys have the same clothing concerns as girls?
5. What advice would you give to Rhiannon and other teenagers about managing resources to meet their clothing needs?

Obtaining New Clothes

Completing your wardrobe will probably mean adding some clothes and *accessories*. You can choose to do this in several ways. You can shop for new clothes in a variety of places. If you have the skills, you might be able to sew some of the items you need, or perhaps you can purchase the fabric and pattern so that someone else can sew them for you. Some friends trade clothes that they do not want any more to expand each other's wardrobes. Old clothes can be remodeled or altered to make them usable again. Investigating all of your alternatives can save you the limited resources of time and money.

Shopping for Clothes

If you have decided that you need to shop for clothing and accessories, begin by doing some research. You could start by looking at advertisements for the items you need. You can determine the latest styles and colours, where the clothes you like are available, and the prices. You might prefer to start by comparison shopping at different stores. Compare the styles and prices at specialty stores, department stores, and discount stores in your community. You might be able to buy basic items like T-shirts, jeans, and athletic wear for less at a factory outlet if there is one in your area. When you find the styles and colours you want at the price you can afford, check the quality and try them on to check the fit.

Shopping by mail, phone, or the Internet can be a faster alternative for some garments. You need to do some homework before you order. You will need someone's help to take your body measurements accurately. Compare your measurements carefully with the sizes available. You will usually be able to return clothes that do not fit or exchange them, but doing that eliminates any time savings. Read the description of the garments carefully. If you do not know the meanings of the fabric names or the special features look them up or ask questions. Read the fine print about returns and do not forget to add any handling charges to your payment.

Recycled clothing is another option. Buying clothing from charity or consignment stores enables you to buy a bargain and protect the environment—as long as you only buy bargains. Used clothing should be examined closely to ensure that the fabric is not worn and that the seams, hems, and closures are still strong. Avoid stains that may be permanent by now. Try on the garment to make sure it fits. Decide how much wear the garment has left. Carefully consider whether the price is a good value and be prepared to pay more for trendy "vintage" garments.

Sewing Clothing

The skills needed for sewing clothes can be valuable resources for you, now and in the future. Once you have the basic skills for selecting a pattern, choosing fabrics, taking measurements, and following instructions, you will be able to sew and repair clothes for yourself and your family. You will be able

> *Socks—life's small luxuries—are supposed to be personal and it doesn't get much more so. By stocking up on them, we can have the pleasure of knowing that the only holes are leg holes—go head, step in front of a bus!*
>
> Jim Gregor

Quality that Fits!

Examine garments before you try them on and then while you are wearing them:

Examine the label.
Check the care label. Clothes that will be worn frequently should be washable, preferably by machine.

Examine the fabric.
Crumple the fabric and release to check for wrinkling. Feel the texture of the fabric. Unless the wrinkled look is in fashion, a fabric that resists wrinkling looks neater, and will require less ironing.

Examine the seams, hems, and closures.
By pulling gently on the stitching you can check whether they are well sewn and durable. Buttons should be large enough to fit securely in their buttonholes. Zippers should open and close smoothly. When you have the garment on, check that the seams and hems are smooth with no puckering.

Examine the fit.
Clothing sizes are not standardized so try more than one size to get the best fit. Do up the closures and check that the garment hangs comfortably on your body. Girls should check the placement of buttons over the bust to avoid gapping. Check that the garment does not bind in the armholes, waist, or crotch areas when you bend, sit, and move your arms. You might have to step out of the dressing room to check for freedom of movement for active wear.

Understanding the vocabulary of sewing will enable you to follow the pattern instructions to sew a garment successfully.

to add special details to your clothes to make them unique. Using your skills for sewing some of your clothes saves money. To acquire the skills you need, learn and practice them. Sewing books, clothing courses at school, workshops at home-sewing stores, and the Internet are good resources for learning how to use patterns and developing your sewing skills.

Patterns are used for sewing. If you can follow instructions, measure accurately, and have a basic knowledge of sewing terms, you can use a pattern to make your own clothing. Patterns are available at home sewing stores. Magazines are another common source of patterns. A new way to get patterns is to order unique designs from a designer's Web site or to custom fit a pattern using computer software. Begin to develop your sewing skills by choosing a basic pattern with detailed instructions.

Sewing patterns contain several parts to ensure that the sewing project will be a success. A list of materials required is included so that you can shop for the fabric, thread, zippers, and any other notions you do not have on hand before you start sewing. Most patterns contain both metric—the Canadian standard—and imperial measurements, but they are not identical so pick a set you will use and stick with it. Pattern layouts are diagrams showing how to place the pattern pieces on the fabric so that you can cut out the fabric pieces you need for the garment. Illustrated step-by-step instructions explain how to sew the pieces together to make the garment. Check the meaning of the terms, such as fold, baste, stitch, finish, or ease, and follow them precisely. You will have much better success if you remember to use your iron to press the seams and details as you finish each step. Practise following patterns carefully and you will be able to sew the exciting clothes you see in magazines.

CASE STUDY

Lisa's Story

My name is Lisa Ho and I am a student at Holy Name of Mary Secondary School. I am taking a clothing course as one of my electives. I find great interest in this course because it gives me a chance to get creative and learn at the same time.

I have been able to create wonderful garments that are fashioned and fitted to my own size and preference. For my independent study unit, I was fortunate to choose what I desired to sew. I am very fascinated with traditional clothing and that's why I chose to sew a dress from the 1600s. This garment was intended to be worn as a Halloween costume since this type of costume is uncommon. That's why I love this course. It helps me learn the techniques of sewing as well as techniques to customize the garment.

The best thing about learning to sew is that you could save a lot of money. If I were to find a dress downtown like the one I made, it would cost me more than $300 to buy. I only spent $45! Overall, I worked for 16 classes at school and some time at home. I used the sewing machines and sergers that were available at school and at home. I learned the sewing techniques from my teachers and my mother. Because I am so petite, it is very hard to find clothes

that actually fit me, but by making my own clothes, I can avoid major alterations. I also save money by not buying commercialized clothing brand names.

I feel that sewing is truly a talent and a skill that most people should try out because it is fun and adds to everyday life. Doing this project helped me with my time management and strengthened my focus. I love working hands-on with fabric and a sewing machine. I love to express my creativity and it gives me confidence to try new and challenging things. The ability to work independently is what makes this hobby so stimulating. What I like most about this course is that I get the privilege of using my creativity, preferences, and knowledge to produce a garment that is useful. Just having the satisfaction of making something on my own rather than buying it in the store is a good enough reason to pursue sewing as a hobby in the future.

What do you think?

1. Have you ever made anything that you could wear using sewing, knitting, or other skills?
2. What skills can you develop by learning to make your own clothing?
3. Why was Lisa able to save so much money by making her Halloween costume?

✓ List the types of skills needed for sewing and repairing clothing.

✓ Why are sewing skills valuable resources?

✓ Why are patterns used for sewing clothing?

✓ What information is provided in patterns to enable you to sew a garment?

Technology can be used to complete some of the steps more efficiently. It is no longer practical to sew an entire garment by hand, although repairs can be made. A sewing machine or a serger substitutes money for the time and energy required to complete the task by hand. Other machines from electric scissors to computerized embroidery machines are available. When deciding whether any new technology would be a good purchase, consider the time and energy that will be saved relative to the cost. Think about whether the technology is convenient enough to use regularly. Carefully selected, technology can provide opportunities for you to express your creativity in new ways and have fun while you sew the clothing you need.

Recycling and Repairing Clothes

What do you have hiding in your closet? Do you have clothes that you do not wear because they need repairs? Are there clothes that you like enough to keep but do not wear because they are too long or too wide? Are there some garments that need something to make them more exciting? Making the necessary improvements to your clothes can save you the cost of replacing them and reduces waste. Begin by assessing exactly what needs to be done to make each garment wearable. Gather the scissors, pins, thread and needle, or sewing machine and you are ready to begin.

When repairing clothes, attempt to duplicate the original construction for best results. Most repairs can be done by hand.

A natural talent may create the ability to gain material resources. A talent for designing clothes could lead to a successful career in the fashion industry.

- Sew seams that have pulled apart by sewing along the stitching line with a backstitch, which loops back on itself to create a strong seam that will not come undone, or stitch by machine.
- Re-stitch hems using a durable catch stitch or invisible hem stitch, and take two stitches in the same place occasionally to reinforce the hem.
- If a button has fallen off, garments often have spare buttons sewn to an inside seam, or you might have to replace all of the buttons to get a matching set. Sew the buttons securely and remember to lock the thread at the back with several stitches in place. Restitch loose buttons at the same time.
- Small tears can be repaired by sewing a patch to the inside of the garment behind the tear. A faster method is to purchase an iron-on patch at the fabric store and attach it to the inside. When you are making repairs, check for other weak details at the same time and repair them also to save work later.
- Making alterations requires some understanding of how to make garments. To take in a garment try on the garment inside out. With someone's help,

DRESSED TO THRILL
These Young Designers Have the Avant-Garde Market All Sewn Up

By Sally Johnston

They are the young and the restless of the local fashion scene.

A small but growing group of underground Edmonton designers has rebelled against the uniformity of chain-store clothing and is producing one-of-a-kind garments and accessories.

Their work is diverse—ranging from restored 1950s prom dresses and metallic Goth chokers to red faux fur pants and satin corsets.

"There is so much talent in Edmonton," said Cherie Howard, the new owner of Sputnik boutique in downtown's Boardwalk Market. Howard, a 21-year-old former hairdresser, bought the store four months ago and has turned it into an outlet for budding jewelry and clothes designers.

"It's great that someone young and hip is bringing the focus on alternative designers to the downtown area and not just Whyte Avenue," said Nicole Reeves, who makes silver and copper brooches and necklaces.

Unlike Reeves, who has an arts degree and a jewelry studio, many of the 16 youngsters who consign to Howard have no formal fashion training and are home-based. They don't advertise. Most started making clothes for themselves and get orders by word of mouth.

"Mass-produced fashion is very drab and unimaginative." said Michal Wawrykowicz, who makes jackets he describes as avant-garde. The bespectacled 18-year-old Centre High school student says he likes to "manipulate things that already exist." In other words, he cuts up old clothes and, using his mom's sewing machine, pieces together the scraps to create a brand new look.

One jacket has a telephone keypad stitched in place of a pocket. "It

Howard models her wild Cherry Pepsi dress.

represents how technology is seeping into our lives," he said.

Wawrykowicz, like many of his young counterparts, taught himself how to cut fabric and sew. A few, like Cori Giacomazzi, who specializes in corsetry to be worn as both outerwear and underwear, have taken textiles and clothing courses. What the designers have in common is a bold originality that is always fun and sometimes artistic.

"I got tired of spending a lot of money on clothes that were not what I wanted," said Sam Hodge, 19, explaining how she started her label, Blue, a year ago. "If I am going to spend $100 on something to wear, I want it to be unique," said the orange-haired young woman. Her bright-coloured overalls and baggy pants, which sell for around $125, have been a hit with friends. She's already completed about 15 orders as word

of her talent spreads. Hodge, who works by day as a landscaper, plans to take art and design courses at Grant MacEwan College in the fall.

Mieke Leonard, 18, fixes up vintage clothes she buys through Internet auctions. She repairs tears and replaces tattered lace, beads and rhinestones. The sociology student has no plans to make a living from stitching but regards it as an enjoyable hobby.

Chris Ethier, however, is planning to go online with his leather and stud collars, cuffs and other bondage-style gear. "I'm building my Web site and will have it up and running by the end of October," said the 22-year-old music store employee.

Howard's own creations, under the name Cherry Bomb Designs, include a dress composed of nothing but 40 Wild Cherry Pepsi cans, flattened and shaped into triangles and held together with 200 safety pins. She wears it over a black sheath slip. "Clothing is not just to cover the body," she said, "but to express who you are and who you want to be.

"I love Wild Cherry Pepsi."

The Edmonton Sun, 06-11-2000

Mieke Leonard wears one of the vintage dresses she has restored.

pin the new seam lines, balancing both sides of the garment or each leg. Restitch the seams in a smooth line from the original line to the hemline. Trim the seam the same width as the original seams and finish the edge. Restitch the hem. To shorten pants, skirts or dresses, try on the garment and have someone mark the new hemline by measuring up from the floor all around the hem. Examine the old hem to determine how wide the hem should be and how you should sew the hem. You might have to unpick the old hem and press it flat before you cut off the extra fabric to leave the hem allowance. Turn up the hem allowance and stitch the new hem to duplicate the old one. Finish your alterations by pressing the garment.

Some garments can be recycled into new garments by unpicking the seams and hems and cutting into a new garment using a pattern. This is also the method for making major alterations if the garment is too big all over. You can even take apart a comfortable but worn out garment to use as a pattern. Reuse the original buttons and buttonholes, collars, and neckline, if you can, to save work on these more challenging features. Experiment with some of the decorative ideas. Patches, embroidery, and trims are details that you can add yourself to achieve a current fad. You could get a new garment by using your skills and your time instead of using money.

CHECK Points

✓ Why is it good idea to know how to make basic repairs to clothes?

✓ What are the most common alterations teenagers make to their clothes?

✓ What features would be fashionable to add to clothes to bring them up-to-date?

Reflections and Connections

1. What sewing skills do you already have? What skills do you want to acquire next?

2. What have you done with garments that needed repairs in the past? How much did the solution cost?

3. What supplies should you have on hand in order to be able to make the most common repairs efficiently? Do you have them in your home?

Hands On!

1. Interview students to find out their most common repairs and compile the results. Make the repair yourself, recording each step using a camera. Prepare a flyer showing how to make the repair, illustrated with step-by-step photographs, to be shared with other students.

2. Select a garment that needs to be altered. Using the resources available in your home, your school, and your community, investigate how to make the alterations. Make the alterations. Identify and apply criteria to evaluate the benefits of making the alterations instead of buying a new garment.

Summary

1. Clothing is designed using the elements and principles of design to enhance the appearance of the wearer.

2. Clothing choices can reflect the personality, interests, and values of the wearer.

3. Wearing appropriate clothing for the social, cultural, or work situation will improve social interaction.

4. Wardrobe planning should consider the appearance, personality, and activities of the wearer as well as the resources available.

5. When purchasing new clothing from retail stores, or used clothing from second-hand outlets, examine the quality and fit carefully.

6. Before ordering clothes by phone, mail, or on the Internet, investigate the quality, sizing, and exchange policy.

7. Developing basic sewing skills can enable you to sew your own clothing and make basic repairs or alterations in order to save money and express your own style.

In this chapter, you have learned to

- apply decision-making skills to choosing clothing that meets your needs and goals;
- plan a wardrobe that enables you to participate in the activities of your life;
- identify and demonstrate the practical skills for clothing acquisition and care; and
- evaluate the information, advice, and technology available as resources for making clothing decisions.

Activities to Demonstrate Your Learning

1. Working in collaborative groups, develop a wardrobe plan for a student who will be working at a summer job where the clothing requirements are different from those at school. Investigate the dress code for the job. Determine what garments and accessories will need to be added to a basic wardrobe, considering how often clothes can be worn, how clothes can be combined into various outfits, and how often laundry can be done. Determine how the garments can be obtained in your community within a realistic budget and make a plan.

2. Conduct a gap analysis of your own wardrobe. Identify your own clothing requirements. Do an inventory of your clothes to determine which clothes are wearable, which need repairs or alterations, and which should be disposed of. Summarize the clothing that you need to acquire and develop an action plan that will make the best use of your resources. Complete one action from your plan and report on how you managed your resources to achieve the goal.

Enrichment

1. Select a pattern and fabric or a garment which you can recycle, and sew a simple garment for yourself or someone else.

2. Organize a class project to use basic sewing skills to provide a service to your community.

Housing

"This is my space and I like it like this." "Home is where the heart is." "My home is my castle." There are many phrases describing living space. How do you feel about the space you call your own? What kind of home do you predict you will have as an adult? How will technology change the home of the future? Why are homes so different around the world? What impact does one's home have on the global environment?

By the end of this chapter, you will be able to

* apply decision-making skills to using housing to meet your physical, emotional, and social needs;

* summarize practical skills required for home maintenance on a daily, weekly, and seasonal basis;

* describe strategies for acquiring, increasing, and substituting resources for meeting needs within the home; and

* evaluate home safety in terms of fire and other hazards.

You will also use these important terms:

home	non-renewable resources
homeless	renewable resources
household	shelter
natural environment	smoke detectors

Your Need for Shelter

Have you ever wondered about your space? How did your room get to look like it does? Do you like your space? How does your space relate or connect to the space used by the rest of your family? If you share your room with siblings, how is the space divided and used? How you use the space that is yours within your home typically reflects who you are and your priorities, and is affected by the environment in which you live.

One of your basic needs is protection. In Canada, that means *shelter* from the extremes of cold, heat, and rain. The type of shelter that you have depends on many factors. Geographic location is only one factor. The community you live in, whether it is urban or rural, will determine what type of housing is available within your family's price range. Your community can also help to provide you with other basic needs such as protection and safety. Community services such as policing, public utilities, and public health agencies can ensure that you are protected from harm and safe from disease.

The space you call *home* can also meet some of your emotional needs. Whether it is a big or small, modest or extravagant, your home can provide a place where you can be assured of privacy and a peaceful environment. It can be the warm surroundings that provide you with a sense of comfort and security. The comforts of home are often considered to be the warm and pleasant surroundings that are familiar to you. Your home can give you a safe place to be alone in your own space. Your living space can also provide you with a canvas for personal expression. You could decorate your room to make a definite statement about your personality.

Personal space can provide an environment that is comfortable where you can be alone.

Homelessness

In Canada the number of *homeless* people is increasing. More and more individuals and families have no permanent form of housing and live either on the street or in temporary or emergency shelters. Although we know that homelessness is increasing, it is difficult to count the homeless.

Homelessness is a contentious issue and opinions about it are diverse. In a public opinion survey taken in March 1998 by Environics, National Housing Research Council (NHRC), the following results were recorded (March 1998).

1. Organizations like food banks and temporary shelters are suffcent to handle the problem of homelessnesss. (March 2000)

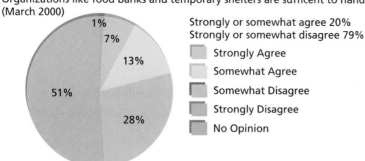

Strongly or somewhat agree 20%
Strongly or somewhat disagree 79%

- Strongly Agree
- Somewhat Agree
- Somewhat Disagree
- Strongly Disagree
- No Opinion

2. The homeless include more than just people living on the street and in shelters. They also include people who must "double up" with others because they cannot find accommodation. (March 2000)

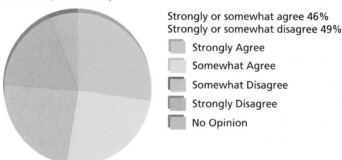

Strongly or somewhat agree 77%
Strongly or somewhat disagree 21%

- Strongly Agree
- Somewhat Agree
- Somewhat Disagree
- Strongly Disagree
- No Opinion

3. Homelessness is a local problem for each community to deal with, rather than the provincial and federal governments. (March 1998)

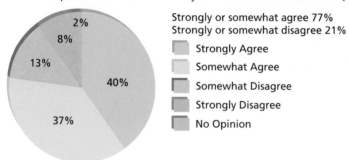

Strongly or somewhat agree 46%
Strongly or somewhat disagree 49%

- Strongly Agree
- Somewhat Agree
- Somewhat Disagree
- Strongly Disagree
- No Opinion

4. The homeless population in Canada is changing and now includes more young people, women and children than used to be the case. (March 2000)

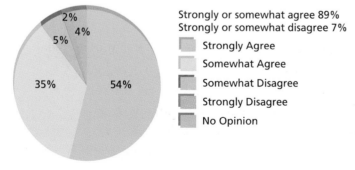

Strongly or somewhat agree 89%
Strongly or somewhat disagree 7%

- Strongly Agree
- Somewhat Agree
- Somewhat Disagree
- Strongly Disagree
- No Opinion

5. I feel uncomfortable when I encounter homeless people. (March 1997)

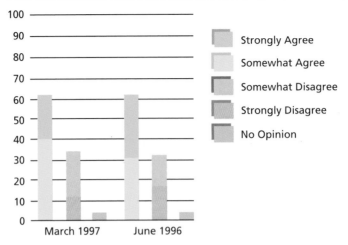

Strongly Agree
Somewhat Agree
Somewhat Disagree
Strongly Disagree
No Opinion

March 1997   June 1996

6. Most homeless people are homeless by choice. (March 1997)

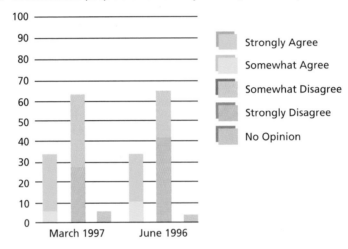

Strongly Agree
Somewhat Agree
Somewhat Disagree
Strongly Disagree
No Opinion

March 1997   June 1996

Canadian Mortgage & Housing Corporation (CMHC)

Reprinted with permission by CMHC

Your home can provide a comfortable space to socialize with your friends and family.

What do you think?

1. What do you know about homelessness in your community?
2. What are your opinions about this social issue?
3. Are you aware of resources for the homeless in your community?

Meeting Your Housing Needs

Do you sometimes invite friends to your home? If you do, then your home is also helping to meet your social needs. You could be sitting around a kitchen table, or on the couch in the living room and enjoying activities such as talking, playing games, or sharing a meal. Living spaces are often designed to enable people to interact with each other. How your home is arranged says something about how the members of your family interact and how others are invited to join in.

Your family may value open space or efficient living. Certain family values are reflected in the location of and the type of housing that they choose.

Where you and your family live is closely related to the resources, values, and goals of the family. Many families have a goal of financial security and choose to live in urban areas where they are more likely to get employment. However, if your family wants to live in a particular city or neighbourhood that is quite expensive, then they may only be able to afford a small apartment. If they want to live in a rural area, where there is less employment, housing costs are often less, but they may be able to afford a larger home. Some families choose neighbourhoods that are close to others of the same cultural background. These communities often provide them with easy access to familiar food markets and cultural amenities. Families often must assess their resources and their goals in order to make decisions about where to live.

CHECK Points

✓ What factors in housing contribute to your physical development?

✓ How does your housing affect your emotional and social development?

✓ Which factors affect the size of home a family may need?

✓ How are a family's goals related to where they live?

Reflections and Connections

1. What determines the type of shelter your family needs?
2. What is the most important place in your home? Why?
3. What would your ideal home be like?

1. Conduct a survey to find out how teenagers in your community use their rooms. Investigate whether they share a room, whether they study in their rooms, use televisions or stereos, etc. Based on the results of your investigations, determine the concerns about improving the use of housing space for teenagers.

2. Take a look at your space, either at home or at school. Identify the materials that have been used to construct and enhance the environment. For every material used, list the reasons for its use and give examples of some other materials that could have been used in its place.

3. Design a room of your own which reflects your self-concept. Use the furniture and other resources available in your home as much as possible. Using catalogues and advertisements, calculate the cost of the room and make adjustments to fit your family resources. Perhaps you can make this dream room a reality soon.

Your Space or Mine?

Families divide their living spaces into shared space and private space. Areas such as kitchens and living rooms are the most used common areas. These are also the spaces that are used for socializing and are often visited by friends and relatives. Social spaces such as these are often decorated to give messages about the values of the family. Bedrooms and, sometimes, bathrooms are usually private spaces. These rooms can be decorated to express the personal values and preferences of the individuals that use them. Most importantly, living spaces should be designed to provide comfortable places for everyday activities for the people who use them.

Social spaces are sometimes used to demonstrate a cultural interest or to display meaningful possessions.

Some families demonstrate that they value friends and socializing by creating spaces that are comfortable for entertaining friends.

Designing Your Space

Living spaces provide opportunities for creative expression. Family members can use the elements and principles of design to create rooms that convey different messages. A home is a place where the family can express their tastes, culture, and personality through design.

Colour is the element that can have the most impact on the atmosphere of a room. Warm colours such as orange, red, and yellow can create a feeling of warmth. When combined with rich wood tones, warm colours can make a room cozy. Cool colours such as green and blue can create an atmosphere of serenity and calmness. Cool colours can create spaces that invite rest and relaxation. Hospitals often paint patient rooms in cool colours to create an atmosphere of peacefulness and calm. The colour of a room significantly affects the way that people feel within its environment.

Line refers to obvious lines within an object or the outline of an object. Lines can cause the eye to move in certain directions and create optical illusions of height or width. Lines can be used to make rooms or furnishings appear larger or smaller. For example, vertically striped wallpaper can make a room seem higher than it is, while horizontally striped wallpaper may make it look lower and longer.

Texture refers to the feel of an object when it is touched. Wood, glass, and ceramics are hard and smooth; these textures appear clean and modern.

What message does the use of colour in this room communicate?

Soft and plush texture can provide softness to furnishings to create a mood of comfort. Texture can reflect light if it is shiny, or absorb light if it is rough. The textures that are chosen for a room express the interests and the style of the residents, and can create a mood that reflects how the room will be used.

Form and shape refer to any object's three-dimensional shape. An object's form is determined by its function, or how it will be used. For example, a chair is designed to allow you to sit in a comfortable position with your back upright and your feet on the floor, and a table must have a stable flat surface. Well-designed furniture is functional and has good form so that it is attractive to look at, also. Rooms also have form that can be used to create a sense of spaciousness or coziness by the clever use of colour, line, and texture.

How the elements of design are used in a room can determine whether the room is comfortable and practical. Colour, line, texture, and form may be applied in many ways to communicate a variety of effects. How the elements of design are used and what effect they have on the eye is determined by principles of design.

You can emphasize a focal point or area of interest in any space. Some rooms may have a natural point of interest, such as a large window or a fireplace. In others, you have the opportunity to create a focal point by applying the principle of emphasis. Using colour is an effective way to create emphasis.

Balance can be formal or informal. Formal balance can be achieved by arranging furnishings in equal amounts or distances from a centre point. Informal balance occurs when items are purposely arranged so that one side seems to have more weight. The principle of balance is often achieved through the use of form or shape.

The elements of design can be used to create movement. This sense of movement occurs when one of the elements of design is repeated or is arranged in a progression and causes the eye to move from one part of a room to another.

Sharing Your Space

The space that you and your family live in is a reflection of the values that you hold as well as the relationships that you have. Your room may be quite different from the rest of the family home. You may choose to decorate in such a way as to reflect your personal values and interests. Perhaps you have posters of your favourite musicians or actors, or pictures of things that you find interesting such as cars, sports events, or hobbies. Your room can also be your personal museum where you display the mementos of your life—the photographs, medals, and certificates which reflect your experiences. Your room, whether you share it or not, is the space that you personalize.

Right The arrangement of shapes in this room creates informal balance. The room communicates a sense of fun and invites guests to relax and enjoy themselves. **Bottom** Having identically shaped mirrors creates formal balance.

Sometimes space issues can cause conflicts. Parents may not be comfortable with the way that an adolescent's room is decorated or its level of order and cleanliness. Issues may also arise regarding "auditory" space. If one person in the family is listening to music or watching television at a volume that disturbs other family members, problems may arise. In these types of situations, skills in communication as well as decision making may help to reduce or resolve conflict.

CASE STUDY

Amanda's Story

I love to listen to music in my room. I like it loud! This is a big problem with my parents. They are always telling me to turn it down, but then I feel that I am not really able to get into my music. If I am looking at magazines or just relaxing I can wear my headphones because I really do not want to *annoy my folks. But a lot of the time, I am doing exercises and stretches in my room—I can't wear my headphones! This is a big problem with my parents!*

What do you think?

1. How might Amanda resolve this issue with her parents?
2. What technological solutions might she use?
3. What communication skills would help to resolve this difficulty?
4. Outline how she might work through a decision-making model with her parents to come up with a satisfactory solution.

CHECK/Points

✓ What effect can colour have on emotions?

✓ How can line be used to create an optical illusion?

✓ What is the difference between formal and informal balance?

Reflections and Connections

1. What values do you demonstrate by the way you decorate and keep your room?

2. If you were to have a conflict about space with one of your family members, what strategies could you use the resolve the conflict?

Hands On!

1. **Determine the noise level of various areas of your home. Describe the activities that occur in each of these areas and the sources of noise. How do the noise levels enhance or detract from the activities that are carried on in each area of your home? Make a list of recommendations that would enhance communication and other activities in each specified area.**

2. **Question your classmates. Ask them to describe three or four items that they have on display in their room and why they are important enough to be displayed. Identify the values that each of the items represents.**

Wheelchair access or other modifications to assist with mobility might be necessary.

Housing in the Global Village

Homes around the world have many things in common. They all provide a place within which to meet the physical, emotional, and social needs of families. Homes have public spaces for receiving visitors and socializing in groups. They also include private spaces for sleep and personal care. Homes require spaces for food preparation and laundry. Some homes include workspaces for earning an income. The way homes look, how they are organized, how they are built, and the ways that families use their homes differ greatly around the world.

Your family's size, form, and stage within the life cycle are also determining factors in the type of living space you choose. Single people and couples do not always need as much space as growing and extended families. Special needs of family members might also determine the type of housing selected. Families consider all of their needs when choosing their living spaces.

The structure and form of housing is greatly dependent upon geography and climate. The type of home required in the Amazon jungle is very different from the shelter that is necessary for a family in the Arctic. In colder climates where shelter from the cold is important, materials that can withstand extreme temperature fluctuations are used. The size of home and the materials used is often related to the amount and types of materials available.

Culture and tradition affect the style of a home. One example is the use of open and closed spaces. In Canada we have a relatively open use of space. Our front yards are open to the community and large "picture windows" provide us with a larger view of the outside community. Many living spaces flow into one another for multiple use, such as the living-dining room. Some people, on the other hand, like to keep their personal space very private. Inside they use double doors for soundproofing, and outside they tend towards high fences so that their yard is not within the view of their neighbours. The spaces in Middle-Eastern homes, and some other warm climates although open and airy, are designed specifically to draw family members together, while keeping out the eyes of the public. Throughout the world, cultures have developed building styles that are unique to them, based on folklore, custom, and lifestyles.

CHECK Points

✓ What determines the size of home and materials used in different parts of the world?

✓ How can technology help to bring the outdoors into our homes?

Below: Many homes in Canada are designed to include the outdoors as living space. Verandahs, such as the one below, are often used for social and recreational purposes.

Geography and climate determine what materials are available for housing. Traditionally, homes have been made from locally available materials. Today, with easier access to transportation, a variety of materials are available in almost any location.

In Canada, when the weather allows, the outdoors is often extended into living space. Balconies, porches, decks, and backyards are used for social and recreational purposes. Various uses of technology have also allowed us to bring the outside indoors. Skylights, greenhouse windows, and solar energy panels are a few examples of how we can incorporate the benefits of nature into our homes.

Reflections and Connections

1. What cultural influences can be seen in homes in your city or neighbourhood?
2. Look around at the homes and buildings in your neighbourhood. Why do you think they are made from the materials that they are made from?
3. How does the design of your home reflect Canadian cultural influences in the use of living spaces? Has your family made changes in how the space is used to reflect your culture?

Keeping It Clean—Housework Never Ends!

A major responsibility of families is providing the necessities for family members. Maintaining the family's *household* is part of this responsibility. Housework involves tasks basic to everyday life, ranging from making beds, to preparing meals, to taking out the garbage.

Caring for your home environment has many benefits. It is easier to live in a tidy space than in a messy one. You can find things more easily when they are in their proper place. Taking care of things makes them last longer and stay in better condition. Food-born illnesses and viral infections can be transmitted by unsanitary conditions, so a clean home can reduce illness. Household pests such as mice and roaches tend to be more of a problem in unkempt conditions. A messy space can also make you feel unhappy, pressured, or depressed because your environment influences your state of mind. A clean and tidy home can make you feel comfortable and relaxed.

Dirt and harmful bacteria can lead to illness. Clean environments are not as likely to transmit germs.

Maintaining a clean and healthy home is easier than cleaning up a big mess. It is quicker and easier to clean as you go. Household jobs can be organized into three time patterns: tasks that need to be done daily, those that are done about once a week, and those that are done seasonally. Although families use a variety of strategies for dividing up the responsibilities, you can take on more responsibilities as you become more mature and accept a larger role in your family. You can also learn how to perform the more skilled household tasks in preparation for living independently. You can manage your time so that you can contribute to household tasks on a regular basis.

Time Categories for Household Tasks

Daily Tasks
- food preparation and clean-up
- cleaning personal care areas
- organizing bed, dressers, and closet
- straightening personal and family possessions such as books, CDs, and sports equipment

Weekly Tasks
- shopping and putting groceries away
- sorting and washing laundry
- collecting and disposing of garbage
- sorting and putting out recyclable materials
- vacuuming and dusting
- mopping kitchen and bathroom floors

Seasonal Tasks
- cleaning closets
- cleaning windows
- polishing wood surfaces such as floors and furniture
- cleaning the oven
- checking batteries in smoke detectors and carbon monoxide detectors

Working together can make daily tasks go more quickly. Work that can be done with others can also be more fun.

CHECK Points

✓ Why is it a good idea to do house-cleaning on a regular basis?

✓ How can resources be substituted in the housecleaning process?

✓ Why is it important to incorporate safety precautions into regular household tasks?

Reflections and Connections

1. What housework do you do?
2. How are the housework tasks divided in your family?
3. What are the most popular and least popular tasks in your family? How do your preferences compare to those of your classmates?
4. What tasks could you take on, as you become more responsible within your family?
5. What housework tasks do you need to learn to do before you become independent?

Hands On!

1. Make a list of the physical, emotional, and social needs that families help their members meet. For each of the needs, identify an area, a room, or some furnishings that help to meet the specific need. Specify areas that meet many needs for family members. Summarize your information in the form of a poster.

2. Arrange a class debate on the resolution: Be it resolved that teenagers should be free to decide on the appropriate level of housekeeping for their own room.

Managing Family Resources

Meeting Safety and Security Needs

Housing meets the basic need for shelter and safety for families. Although a home, whether it is a small apartment or a spacious mansion, provides shelter from the environment, it is also the place where most accidents happen. You can contribute to your family's safety by being aware of potential dangers in the home.

Falls

Falls are the most common accident that can occur in your home. Small children and older adults are most likely to suffer from accidental falls. Most falls can be avoided by careful strategies that reduce the risk:

- install safety gates at the top and bottom of stairs
- keep staircases, hallways, and doorways clear
- secure handrails on all staircases
- anchor throw rugs with carpet tape or non-skid mats
- secure electrical cords to baseboards
- use step ladders, instead of chairs or stools, to reach high places
- ensure adequate lighting so that people can see where they are going

Fires

Fires are among the most serious home accidents. Fires can cause disabling and disfiguring burns, and they are often fatal. Most fatal fires occur at night when family members are asleep. Smoke kills people more frequently than the actual flames of a fire. Most home fires are avoidable.

Smoke detectors should be installed in your home and are required in most Canadian homes by law. A **smoke detector** is a device that sounds an alarm when it senses smoke. Since many fires occur at night, a smoke detector may save lives by waking people up and alerting them to the fire. They should be installed in every level of a home. Most smoke detectors run on batteries, which should be changed every six months. Smoke detectors can reduce the risk of death or injury from fires.

Another resource for preventing loss from fires is a fire extinguisher. Your home should have a fire extinguisher in the kitchen. Place it in clear sight near the exit from the kitchen so that it can be grabbed quickly in the event of a small fire.

Fire Prevention

- Store items away from a heater, furnace, or stove.
- Discard or recycle papers, wood, oily rags, and any other flammable refuse.
- Follow manufacturer's instructions when using flammable materials and solvents.
- Make sure matches are cold and wet before throwing them away.
- Keep matches and lighters out of the reach of small children.

Many common household products are potentially lethal when swallowed. Care must be taken to make sure that these products are properly labelled and kept out of the reach of children.

Poisoning

Poisonings occur most often in the kitchen, bathroom, or bedroom. Small children under the age of six are most at risk because of their natural curiosity. Today, most medicines, household cleaners, and other potentially toxic products are packaged in containers that are difficult for children to open. Yet poisonings still occur, often due to human error.

Store potentially toxic products safely. Medicines and household chemicals such as cleaning products should be stored in a safe location away from food. Other items such as glue, varsol, and turpentine should also be kept in a secure place, out of the reach of children. All toxic substances are required by law to be labelled, so never store poisons in another container such as a pop bottle. Even adults could mistake the substance for something that is safe. If poisoning occurs, call a poison control centre immediately. They will tell you the best way to deal with the situation.

When there are children living in or visiting the home, extra precautions are necessary. Medicines should be stored out of reach, preferably in a locked or childproof cabinet. Household cleaning products and hobby supplies should also be stored in locked cupboards. Childproof locks are available which can still be opened easily by adult hands. If a member of the family takes medication regularly, extra caution is necessary. Be sure to keep all medication out of the reach of children at all times. Children are naturally curious and some children have even been poisoned with pills from a visitor's purse. Keep the local Poison Control Centre telephone number by your phone for emergency use.

Carbon monoxide is an odourless gas that can be deadly. Carbon monoxide fumes can be produced by furnaces, gas stoves, gas heaters, and other gas appliances that are not vented or working properly. If carbon monoxide poisoning happens while a family is sleeping, they may never wake up. If you have any sort of gas appliance in your home, you should install a carbon monoxide detector. It operates much like a smoke detector. When it detects carbon monoxide, it will sound an alarm. This alarm will alert the family to the problem and potentially save lives.

Electrical Hazards

Electricity is an important resource in your life. It provides a reliable source of energy for many of the things you do. Do not forget that electricity can be very dangerous or even fatal. To avoid electrical hazards:

- to prevent circuit overload, avoid plugging too many electrical appliances into the same outlet;
- to avoid getting a shock, never use an electrical appliance of any sort near water or touch it with wet hands;
- repair or replace damaged electrical cords or appliance parts immediately;
- do not run electrical cords under rugs or carpet, or across the floor in a traffic area; and
- cover all unused outlets with childproof covers when there are small children in the house.

Every Area in the Home Has Potential Safety Hazards

Kitchen
- Wipe up all spills immediately.
- Keep stove and sink areas well lit.
- Keep pot handles turned to the centre of the stove.
- Dry your hands before using an electrical appliance.
- Keep a fire extinguisher handy, make sure that it is functional.
- To prevent fires, keep areas around the stove clean and free from grease.

Living Spaces
- In cold climates, keep steps and walks clear of snow and ice.
- Keep traffic areas well lit.
- Arrange furniture out of the path of heavy traffic.
- Keep areas by front and back door clear of any obstructions.

Bathroom
- Never use an electrical appliance while in the bathtub.
- Keep a night light on in the bathroom.
- Keep medicines out of the reach of children.
- Wipe up water to prevent slipping.
- Make sure mats have a non-skid backing.

When Emergencies Happen

Emergencies are never planned, but it is wise to have a plan, in case of an emergency. Being prepared can save precious time and possibly save lives.

1. **Have a family meeting and identify potential hazards and related safety precautions in your home. Identify escape routes and then hold a fire drill.**

2. **Post emergency phone numbers near the phone so that everyone can see them or use the memory dial feature on your telephone:**
   - police, fire, and ambulance (911)
   - poison control centre
   - doctor or clinic
   - the closest relative
   - work phone numbers of parents or guardians
   - phone number of friend or neighbour

3. **Make sure that you have a complete first-aid kit.**
   - adhesive bandages in various shapes and sizes
   - gauze rolls
   - sterile gauze pads
   - adhesive tape
   - antibiotic ointment
   - disinfectant
   - first-aid manual
   - list of emergency telephone numbers

CHECK Points

✓ Why is it important to have smoke detectors on every level of every home?

✓ How can poisonings be avoided?

✓ Why is the kitchen a particularly dangerous place?

Using Space Efficiently

Today's busy families have many things they want to do in their living spaces. Often there are more tasks to be done than rooms available to do them in. Using space for multiple purposes is one way to make the most of the space available. For example, in many families, the kitchen table or the dining room table serves as a work area after mealtime is over. As long as there is adequate lighting and the room is quiet, these spaces provide an excellent area for you to do your homework. Using these areas is particularly useful when you share a bedroom with a younger sibling who has an earlier bedtime.

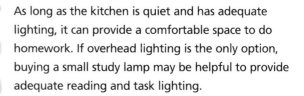

As long as the kitchen is quiet and has adequate lighting, it can provide a comfortable space to do homework. If overhead lighting is the only option, buying a small study lamp may be helpful to provide adequate reading and task lighting.

Incorporating an office into a living room or family room can be as easy as adding one piece of furniture. This self-contained office unit has room for a computer, software, office supplies, and reference materials.

A loft bed creates more floor area in the bedroom. The space under the bed could be used for a home office, music centre, or for storage.

If a family member has a hobby, then the kitchen or dining table can also provide a work area. A portable sewing machine or a large puzzle board could be set up on a large table. When a home office is required, space must be made for storage and equipment. If a specific room is not available, then a portion of a room may serve the purpose. A table and bookshelves can provide the space that is necessary for a home office.

If you need to share a bedroom, there are many ways in which space use can be maximized. Bunk beds are one way of increasing sleeping space without increasing the use of floor space. Even in rooms where sharing is not an issue, raised beds can provide sleeping space and still allow the room to be used for other activities such as exercising or entertaining friends. Bookshelves and other tall furniture such as armoires can be used as room dividers. They can provide storage as well as visual barriers to divide space for different purposes or to create privacy.

Functional Decorating

The way you decorate a room can also make it more functional. If you need more storage space than your furniture provides, some decorating tricks can create more space. The creative use of fabrics can maximize storage space. By using a floor length tablecloth you can create a large space for storage under a table while still having the use of the tabletop. A dust ruffle around the bottom of a bed can hide storage boxes filled with off-season clothing and sports equipment. Colourful fabric bags can be made to hold laundry, shoes, and other items. Fabric can also be used to make storage compartments for reading materials by the bed or a comfortable chair.

A television accessory holder can eliminate the need for a table or drawer to hold the remote control and the television channel guide. Boxes can also be purchased or made to co-ordinate with room decor and provide interesting storage space.

When space must be used for both socializing and sleeping there are a few options. Furniture can be purchased that is multi-functional. A pullout couch provides seating within a social space and a comfortable bed at night. Similarly many futon couches and chairs fulfil this same function. If you already have a bed and cannot afford to buy new furniture, then many pillows can be added to a bed to create a comfortable seating area by day and a bed at night.

Reflections and Connections

1. Think of an area in your room or your home that is not used very much. What could you do to make the area more functional or useful?
2. What are some things that you can do to provide more organized storage in your personal space?

Environmental Concerns about Your Home Environment

The decisions we make on a daily basis in our homes have an impact on the local and global environment. Polluted air and water, overflowing landfill sites, and large amounts of toxic waste are often the result of decisions and actions that we make in our homes. Some natural resources are *renewable resources*; that is, they are or can be replaced. Supplies of fresh air, water, and minerals are produced continuously, unless something interferes with their natural cycles. Other natural resources, such as trees and other plants are renewable as well, but often need the assistance of humans to be replenished. The types of housing we choose to live in, the amount of energy we use, the types of products we purchase—all are subject to environmental concerns.

Non-renewable resources do not replace themselves and cannot be replaced in the ***natural environment*** by humans. Minerals such as iron and copper, and fuels such as oil and gas are available now, but will eventually run out. Although supplies of these resources might last a long time, they might also be used up in the near future. Another non-renewable resource that is quickly being depleted is uninhabited land. Both industrial and residential communities are quickly replacing farmland, rain forest, and wilderness areas. In order to protect our natural resources, we must use strategies of conservation—ways of protecting our resources from any waste or harm.

Treading More Lightly on the Earth by Reducing Your Ecological Footprint

If everyone on Earth lived like the average Canadian, we would need at least four Earths to sustain our lifestyle, and provide all the material and energy we currently use. The land needed to support each individual Canadian's current consumption is three city blocks per person, per year! This is the impact, or footprint, that each Canadian makes on the Earth. The Canadian situation is not unlike that of other citizens of the developed world. As it stands now, 20 percent of the Earth's population—its most wealthy citizens—gobble up 80 percent of the resources available for annual consumption. Our ecological footprints are overwhelming our planet's ability to provide for humans—and for the other species that share the Earth with us.

World Wildlife Fund Canada

Water—don't waste a drop!

Imagine waking up tomorrow morning and not having fresh water available. Think of all the things you could not do. Water is critical for food preparation, cleaning, and personal hygiene. It is vital for us to stay alive. Water is a precious resource. You can help to conserve water in many simple ways:

- Take quick showers instead of baths.
- Turn faucets completely off—small drips can waste many litres of water.
- Run the dishwasher or washing machine with full loads.
- Do not wash dishes under running water—this method can waste about 115 litres of water.
- Water lawns and gardens in the evening hours so that the water has time to seep into the earth before the sun begins the evaporation process.

Conserving Energy

Energy is the power that is used to produce something else. Energy can be used for heating and cooling, lighting, cleaning, food preparation, or entertainment. You commonly use electricity as your energy source; however, oil and gas are also frequently used sources of energy. Energy is used in the production of every item that you use on a daily basis.

Some ways that you can save energy in your home include:

- Turn off lights and appliances such as radios and televisions when you are out of the room.
- Dry your clothes by hanging them on a clothes rack or clothesline whenever possible.
- Instead of turning up the thermostat, put on an extra layer of clothing or a heavy sweater to stay warm.
- Open the curtains on a sunny winter's day to get the natural heat from the sun.
- Close the curtains as soon as the sun sets to keep the heat in.

Reduce, Reuse, and Recycle!

By considering such things as packaging waste, you can avoid using the energy it takes to produce, use, and discard unnecessary packaging. Think ahead when you buy products.

- Avoid buying disposable cameras, plates, cups, and diapers.
- Buy in bulk to avoid the packaging that is used in individual serving sized products.
- Buy your food naked—fresh foods, such as fruits and vegetables, require little or no packaging.
- Purchase environmentally friendly household products.

Reuse items or provide them to someone else to reuse:

- Use refillable containers such as manual pump sprayers instead of aerosol cans.
- Use packages as storage containers once they are empty.
- Compost all organic waste such as food scraps to use on your garden.

Recycle everything that you possibly can. Every item that you recycle is one less thing that will go into landfill. You can help the environment and your family by taking responsibility for recycling in your home. Participate in your municipal recycling program, and encourage your school to recycle in the cafeteria and classrooms.

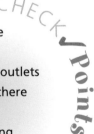

CHECK Points

- ✓ Describe key strategies for fire prevention in the home.
- ✓ Why should unused electrical outlets be covered in a home where there are small children?
- ✓ What are the benefits of having multi-use spaces in the home?
- ✓ List some non-renewable resources along with strategies for conserving them.

Reflections and Connections

1. How do you feel about conservation? Do you practise conservation methods on a daily basis?

2. Do you think about issues like "unnecessary packaging" when you purchase an item?

3. How long do you usually take in the shower? Can you reduce the time in order to help conserve Earth's water supply?

4. Think about the fast food that you buy. Could it be made with less processing and packaging? Would it be more ecologically sound to prepare it at home?

Hands On!

1. Draw a simple diagram of your family's living space. Show potential safety hazard areas on the diagram. List ways to make these areas safe.

2. Design a fire escape plan for your family. Make sure to include a fire drill in your plan.

3. Survey your classmates to see what conservation strategies they use in their homes. Compile the results and present your findings in the form of a report.

Summary

1. Housing meets many basic needs including physical needs such as protection from the extremes of weather, emotional needs such as safety and security, and social needs, such as providing a space to share a family meal.

2. The type of home a family lives in is determined by their stage in the family life cycle, their resources, values and goals, as well as any special needs they might have.

3. Homes provide an opportunity for cultural as well as personal expression through the elements and principles of design. Colour, line, shape, and texture are the elements of design that are combined with balance and rhythm, the principles of design, to allow for creative expression.

4. The structure of homes depends on many factors. The materials that homes are made from depend on climate and availability, while the style is often determined by cultural and other influences.

5. Most accidents occur in the home. It is important to be aware of all potential hazards including fires, falls, poisonings, and electrical hazards. Most home accidents can be prevented.

6. It is important to be aware of all potential hazards and enable measures for prevention. If accidents do occur, families need to know how to deal with emergencies.

7. Spaces within homes can be used for many purposes. Many areas can be made multi-functional by the creative use of decorating and furniture arrangement.

8. In order to maintain health and the general well-being of family members, homes must be cleaned and maintained regularly. Household tasks, completed on a daily, weekly, and seasonal basis will ensure that living spaces are well maintained and support a clean environment.

9. Through the use of conservation methods within the home, families can ensure the supply and maintenance of both renewable and non-renewable resources to ensure the sustainability of our natural environment.

In this chapter, you have learned to

- apply decision-making skills to using housing to meet your physical, emotional, and social needs;
- summarize practical skills required for home maintenance on a daily, weekly, and seasonal basis;
- describe strategies for acquiring, increasing, and substituting resources for meeting needs within the home; and
- evaluate home safety in terms of fire and other hazards.

Activities to Demonstrate Your Learning

1. Discuss how housing can meet needs within your family. Find out what needs are being met by your home. Ask your family members to describe which needs are most important and least important to them.

2. Select a style of home that is represented in your neighbourhood. Find out the origins of that style of home and outline why it is suitable for your particular location and climate.

3. In small groups create a requirement list for housing for families at each stage of the family life cycle. Make sure that you include all physical, social, and emotional needs for each stage. Present your findings to the other groups in the form of a chart.

4. Research how the elements of design can be used to meet the social and emotional needs of families. Present your findings in the form of a poster.

5. Using social science research methods, investigate how your classmates deal with issues of space, privacy, and personal expression in their homes. Formulate research questions and conduct surveys. Compile the results using graphic representations in poster format.

6. Design or desktop publish a pamphlet to give detailed information about how to prevent falls, burns, poisonings, and electrical accidents in the home.

Enrichment

1. Investigate the homelessness issue in your community. How many people are homeless? What is being done to address the issue? Outline strategies that you believe would alleviate the problem of homelessness in your community. Present your findings in the form of a report to the community. Invite a community spokesperson to speak at your school.

2. Many hazards in the home are specific to particular age groups. Investigate which hazards are most commonly experienced by specific age groups and outline the reasons why. For each specific hazard and related age group, prepare a detailed plan for accident prevention.

3. Select a commonly used product in the home, such as an appliance. Investigate the components of the ecological footprint and analyse the size of the ecological footprint made by the appliance or product that you have selected.

Pulling It All Together

Do you ever wonder what path your life will take? What life changing experiences will you have? What will you choose as a career? Will you have a family? Where will you live? Will you have many friends? Will you be ready for the challenges that life may present? This book has provided you with some knowledge and skills that will help to guide you on your journey through life. You have examined the progression of growth and development.

You are now aware that every choice and decision that you make will help to determine what your future lifestyle will be. Your lifestyle includes all aspects of your personal, social, and working life. Your social and professional lifestyles are closely linked. They are affected by the work that you will do and the relationships you will have with people both socially and professionally. To have a satisfying lifestyle you need to be aware of your alternatives and make the choices that are best suited to your goals.

By the end of this chapter, you will understand why it is important to

* communicate your knowledge, skills, and values effectively to others;

* apply a variety of problem-solving and decision-making skills to practical issues regarding family functions;

* internalize communication and conflict-resolution strategies necessary in personal, family, and social relationships;

* personalize the application of practical skills in performing daily-living tasks that meet your needs and those of family members; and

* exercise strategies to manage personal and family resources within a changing social and technological environment.

Are You Achieving Your Potential?

Being You, Now and in the Future: Where Do You Go from Here?

Every generation laughs at the old fashions, but follows religiously the new.

Henry David Thoreau

You have examined the process of how you developed into the person that you are. Now that you have more control over your own life, are there aspects of yourself that you would like to change? You are aware of your temperament and your self-concept. Are there things you would like to improve with respect to your self-esteem, your goals, and standards? Now is the time to begin. Being aware of every action and decision that you make allows you to change your thinking and your behaviour patterns.

In Unit 2, you looked at the multitude of factors that have combined to make you who you are. You examined the inner and the outer you. The outer you is seen by the world and is open to interpretation. Whether it is a true reflection of the inner you is something only you will know. Many of the choices that you make can affect how others see you.

The optional courses you select in school, the television programs you watch, as well as the friends you choose all reflect your values as well as your personality. The interesting thing about life is that the inner you can be in conflict over certain issues. While you are deciding about your values, and learning about who you are, you are required to make decisions on a daily basis. A simple example is your clothing.

Select an item of clothing that you are wearing today and ask yourself some of the questions in this mind-map.

What am I wearing today?

PERSONAL CONSIDERATIONS
Am I wearing this because...
- I want to make a fashion statement?
- it is part of my heritage?
- I think it makes me look good?
- it connects me to my family?

RELATIONSHIPS
Am I wearing this because...
- it is in style?
- all my friends are wearing it?
- my boyfriend bought it for me?
- it is environmentally sound?
- it is required by my job/school?

FAMILY CONNECTIONS
Am I wearing this because...
- it shows my culture?
- it is appropriate for my age group?
- it has some historical meaning?
- it shows my family loves me?

RESOURCE FACTORS
Am I wearing this because...
- I got it on sale?
- it is the only thing that was clean?
- it is easy to care for?
- it is comfortable?

Compare your responses with those of your classmates. Discuss the similarities and differences. Be aware of the number of different factors that impact decisions. The clothing that shows the "outer you" may, in fact, be a financial statement or an environmental statement and not a fashion statement at all.

You might think that the clothes you wear do not have a far-reaching significance. The fact is that clothing touches all areas of life, ranging from meeting basic needs to making an ecological footprint.

Relating to Others

You have examined the nature and essence of relationships. You have identified the different kinds of relationships in your life, how to maintain them, and when to end them. You know the importance of communication in relationships. You know the impact of poor communication on relationships. You have also examined your role in both effective and ineffective relationships.

Relationships are critical to fulfilling many of your needs. If you are having trouble in an important relationship, then you do not feel happy and cannot function to your full potential. Having relationships that fulfil your needs enables you to be effective on a personal as well as social level. As an adult, professional relationships also enter into the mix. All these relationships in your life must be balanced in order to be effective.

In today's society, the demands placed on individuals in their careers are increasing. Professional relationships are sometimes so demanding that personal and social relationships might be neglected. Balancing work and family is a controversial issue in the new millennium. While the pace of business increases, the needs of family members and friends are still important. When contemplating your career path, consider a balance in lifestyle choices.

If the primary goal of future workers is to balance work and personal life, what knowledge and skills do they need in order to accomplish this goal? Is working from home the answer? There are many factors that contribute to the issue.

Students Seek Career Development, Personal Growth and a Balanced Lifestyle According to Pricewaterhouse Coopers

Graduating business students around the world are seeking career development and personal growth while expecting to balance work and life responsibilities according to Pricewaterhouse Coopers' second International Student Survey. Students rate career development (56 percent) and personal growth (55 percent) as their first and second life priorities with attaining a balance between work and personal life as their most important career goal.

Though 57 percent of the respondents state that balancing work and personal life is their primary career goal (up from 45 percent two years

ago), they don't believe that this desire competes with their long-term career development and personal growth goals. The question is not whether personal development is more important than career, but rather how these goals can be achieved in tandem.

While students all over the world make balancing work and personal life their top career priority, they are willing to work hard for it. The survey found that on average respondents expect to work at least 47 hours a week at their first jobs.

In addition to balancing career and personal life, 42 percent of the

students rate a company's reputation for offering a good future career reference as "crucial" when choosing their first employer and they anticipate making an average time commitment of four and a half years to that employer. The factors that most frequently would motivate the respondents to extend their anticipated time commitment to their first employers include a higher salary, ability to balance work and life and rapid promotion.

Put yourself in the centre of this mind-map. Examine the issues and values that may affect your future career choices.

What career path will I follow?

PERSONAL CONSIDERATIONS
Will my career choice match my...
- values and goals?
- self-concept?
- personality?
- interests?

RELATIONSHIPS
Will my career choice reflect my...
- commitment and loyalty to others?
- communication skills?
- need for reciprocity?
- skills in negotiation and problem solving?

FAMILY CONNECTIONS
How will my career choice...
- balance with my life cycle concerns?
- allow for multitasking?
- benefit my family?

RESOURCE FACTORS
Will my career choice...
- highlight my skills and talents?
- challenge my time-management skills?
- help me to balance consumer and government issues?
- provide social supports and taxation?
- provide a satisfactory income?

Career choices have far reaching effects, not only on individuals and families, but also on communities and nations. Your career path may take you anywhere within the global village. The choices that you make now will eventually have an impact on your role as a family member in the workplace.

Living in Families

Health is a state of complete physical, mental, and social well-being, and not merely the absence of disease or infirmity.

World Health Organization

There are lots of people who spend so much time watching their health, they haven't got time to enjoy it.

Josh Billings

You have studied the nature and function of families. You know how various forms of families play their role in nurturing and socializing their members. While families are responsible for meeting the needs of their members, governments provide the support through social policies and services to facilitate the work of families. Many factors influence how families make lifestyle choices for their members.

Maintaining the physical, emotional, and social health of its members is one of the primary functions of the family. What knowledge and skills do family members need in order to fulfil this function? How does our cultural background affect our lifestyle decisions? Is the reliance on government supports adequate to meet family needs? There are many factors that contribute to the issue.

Socializing Is Healthy

By Sally Johnston

Now the health news you have been waiting for—hanging out with friends is good for you.

A new Canadian healthy living checklist says that socializing with family and buddies is as important to feeling well as diet and exercise.

"When we get out and meet people it helps reduce stress," said fitness personality Joanne McLeod, who hosts The Body Break television segments and shows with husband Hal Johnson.

"Talking over a problem ... helps you emotionally and makes you feel better. It prevents you bottling up stress, which can adversely affect your health."

McLeod, 41, and Johnson, 43, are in Edmonton today promoting the Manulife National Health-styles Checkup, which they co-authored with Dr. Michael Stones, a leading expert on aging.

The checklist, which can be completed in about 10 minutes, helps Canadians assess eight areas of their life that affect health, allowing them to plan improvements.

As well as obvious factors such as fitness and nutrition, the list emphasizes the benefits of a busy social life, positive attitude, good support network and planning for the future.

Reprinted with permission
from *The Edmonton Sun*

Lifestyle decisions are critical to achieving and maintaining good health. Ask yourself some of the following questions about health and lifestyle.

How can I be healthy?

PERSONAL CONSIDERATIONS
How will my health be affected by...
- my values and goals?
- my heredity?
- environmental factors in my home?

RELATIONSHIPS
How will I deal with health issues...
- requiring difficult communication?
- needing negotiation?
- demanding tolerance?

FAMILY CONNECTIONS
How will I manage health matters concerning...
- family life cycle concerns?
- social supports for family members?
- available technology?

RESOURCE FACTORS
How will my family's health be affected by...
- my skills and talents?
- time management for socializing?
- my skills in multitasking?
- current consumer and government issues?

Talk to your classmates. What health and lifestyle concerns did they identify?

Family health concerns are many and varied. We all need to support family members in maintaining optimal physical and emotional health. Issues related to health are far reaching. Many feel that lifestyle is affected by pollutants in the environment, others contend that technology is the most critical factor. Sedentary lifestyles, highly processed foods, and complex social environments certainly all affect personal and family health factors.

Managing Family Resources

If I were a Brazilian without land or money or the means to feed my children, I would be burning the rain forest too.

Sting

The supreme reality of our time is... the vulnerability of our planet.

John F. Kennedy

We have forgotten how to be good guests, how to walk lightly on the earth as other creatures do.

1972 Only One Earth Conference

You have looked at the various human and non-human resources available to use in order to meet your goals. Each individual and family has a unique set of resources available to them. The resources selected and how they are used have impact not only on the success of your goals, but also on the earth as a whole.

What resources will you have available to you? Will your resources be adequate to meet your needs or will you have to depend on the community and others to provide you with support? What impact will technology have on the way that you use your resources? How will your personal use of resources affect your family, your community, and the Earth?

The use of resources is an important issue. Economic resources have a huge impact on our daily lives, yet our rapid consumption of our resources could rob us of products that are part of strong economies. If everything that you use and consume has far reaching effects, is it not important for you to be aware of all your resource decisions? What will you do if some resources you depend on are no longer available? What personal resources do you have that can help others to meet their needs? The number of factors affecting this issue is immeasurable.

When you think of the environment, you probably think of animals and trees in a natural setting. Consider the room that you are in right now as your environment. Consider all of the resources that have gone into producing that environment. The list can be mind boggling—everything from the skills and knowledge of the architect who designed the building, to the physical energy of the loggers who cut down the trees for lumber, to the computers that helped technicians to formulate the paint, to the natural power of the rivers that are creating the force for the electricity that is lighting the room, to the farmers who feed all the people involved in the process.

Using a commonly consumed resource, create your own mind-map using four areas of focus.

What resources do I have?

PERSONAL CONSIDERATIONS
With respect to this resource...
- what do I value?
- what responsibility do I feel?
- how does it relate to my life?
- what role does it play in my family history?

RELATIONSHIPS
How will my friends and family...
- communicate about this resource?
- work to preserve this resource?
- compromise on the issues?

FAMILY CONNECTIONS
In dealing with resource use, how will my family...
- adjust to consumer decisions?
- adjust their lifestyle to be responsible consumers?
- adapt to social, economic, and technological changes?

RESOURCE FACTORS
When using this resource...
- what skills do I have?
- how will I evaluate the information and technology available to me?
- what will determine how I balance my use of time, energy, and money?

Discuss resources with your classmates. You might be suprised to find that their perspectives may be very different from yours.

As you walk the path of life's journey, remember to

- consider your strengths, values, rights, and responsibilities;
- communicate openly and clearly in order to maintain good relationships;
- examine and respect the impact of social, economic, and environmental change on families and society; and
- make informed decisions when considering all lifestyle choices

and you will be on your way to achieving your potential for living a healthy, happy, and satisfying life.

Summary

1. As you develop an understanding of the inner and outer you, you are able to make adjustments to your self-esteem, your goals, and your standards.

2. Through many interactions you are learning to develop and maintain satisfying relationships. Balancing relationships within the context of other areas of your life will become increasingly important as you develop personal interests and a career.

3. Understanding the nature and function of families will help you to maintain the physical, emotional, and social health of your family members.

4. Acknowledging the various resources available to you and the many implications of their use will help you to be an environmentally aware customer.

Activities to Demonstrate Your Learning

1. Open the time capsule created when you read Chapter 1. Read the letter that you wrote to yourself. Write a response in which you describe three specific characteristics of growth that you have observed in yourself and explain how you are now better prepared for your life journey.

2. Select one current issue and examine how it is affecting the four areas of your life—individual development, relationships, family life, and resource management—as an adolescent. Begin by gathering information from various media and technology. Using the social science research model, design a survey to determine the opinions of teenagers concerning the impact of the issue on their lives. After a thorough analysis of the information, develop a proposal suggesting an action that should be taken concerning the issue, and recommending who should act.

3. Work as a design team to develop a board game that realistically represents the challenges of adolescent life in the four areas of your life—individual development, relationships, family life, and resource management. Be sure to include the factors you have studied in this book that will contribute to success, combining those that are inherited and those that are environmental. Produce a prototype of the game and playing pieces.

4. Create a poster that presents a precise strategy for success in adolescence. Represent in your design the four areas of your life—individual development, relationships, family life, and resource management. Present the poster in a camera-ready format that could be reproduced for display in the community.

Glossary

Abusive relationship—a relationship that causes one physical or emotional harm

Accessories—something worn or carried, such as a hat, shoes, or a purse, that is chosen to set off or complement a dress, suit, etc.

Acquiescence—acceptance or agreement by keeping quiet or by not making objections

Active listening—hearing a message by concentrating on what is being said to improve understanding and retention

Adolescence—the stage of life that begins around puberty; the teenage years (13–19)

Adolescents—individuals in the stage of life called adolescence; teenagers

Adoptive family—a family in which the parents are legal rather than biological parents of the children, but who bring the children up as though they were their biological parents

Advertising—praising the good qualities of a product, etc. to promote sales

Affection—friendly feeling; a warm liking; fondness

Age of majority—the age at which, by law, a person acquires all the rights and responsibilities of being an adult (in Canada, age 19)

Alternatives—choices from among more than two possibilities

Anthropology—the study of humans and their cultures

Appliances—tools, small machines, or other devices used in carrying out specific tasks

Assertive—firm and positive when communicating one's ideas and feelings

Authority—the power to enforce obedience; right to command or act

Bias—a narrow perspective that makes it difficult to judge a particular situation fairly; often based on opinion and not fact

Blended families—families with parents who divorced their first spouse and remarried, forming a family that includes children from one or both previous marriages; also called recombined or reconstituted families

Body language—a form of non-verbal communication; gesture, attitude, and position of the body

Breadwinner family—a family that has a single income earner and a stay-at-home parent

Budget—a plan one makes for spending and saving money from one's income

Bullying—teasing, frightening, or hurting smaller or weaker people

Canada's Food Guide to Healthy Eating—a document produced by the Canadian government that organizes food into four basic groups to help Canadians make healthy food choices

Care label—a tag attached to clothing and other textiles that indicates the universal symbols for laundry care

Child development—the process of a child's physical, emotional, social, and intellectual growth

Childproofing—making free of danger to or from a child

Choices—options when there are many alternatives available

Clans—families of many related extended families living in the same area

Commitment—dedication; willingness to stick to things in the long term

Common-law—denoting a relationship between cohabiting partners; recognized as a marriage but not brought about by a civil or church ceremony

Communication filters—barriers to clear messages; include bias, stereotyping, and ethnocentrism

Communication—the exchange of information between two or more people

Community resources—community organizations and facilities that enhance the quality of life (e.g., recreational facilities, houses of worship, businesses)

Comparison shopping—looking at the same product in several stores to find the best quality at the lowest price

Compromise—settling a dispute by agreeing that the person(s) on each side will give up a part of what he or she demands

Conjugal arrangement—a live-in sexual relationship

Consequences—results of some previous action or occurrence

Consumer—someone who purchases goods and services produced by others

Consumer rights—consumers' entitlement to safety, accurate information, and to complain when interacting in the marketplace

Credit—security on loans that allows consumers to purchase goods or borrow money on the promise to repay it at some future date

Culture—the arts, beliefs, habits, institutions, and other human endeavours considered together as being characteristic of a particular community, people, or nation

Custody—legal guardianship of a minor

Data—facts or information presented in a form suitable for processing to draw conclusions

Deadline—a specific time by which a task must be done

Decision making—choosing between options

Demographic trend—a pattern of change in the behaviour of a population

Development—progress that occurs when skills are co-ordinated into complex behaviours

Developmental tasks—challenges for growth and development at each stage of life

Discipline—the process by which children are guided to behave in acceptable ways

Discrimination—the act or practice of treating a person differently because of prejudice

Divorce—the legal ending of a marriage

Dovetailing—the process of using "downtime" within one activity to accomplish another

Dual-income family—a family where both parents are employed; the most common nuclear family in Canada

Eating disorders—illnesses that cause the afflicted to be preoccupied with food and weight as a coping strategy to deal with deeper psycho-social problems that are too painful or difficult to address; examples include anorexia nervosa, bulimia nervosa, and binge-eating disorder

Economics—the study of the production and distribution of wealth

Elements of design—colour, line, shape, and texture

Empathy—understanding someone else's feelings or point of view

Employment—work; what a person does for a living

Environment—one's surroundings

Extended families—families that consist of parents, children, aunts, uncles, grandparents, and other blood relations, living together or not

Facts—ideas supported by evidence that can be observed

Family—"any group of two or more persons, who are bound together over time by ties of blood or mutual consent, birth and/or adoption and who, together, assume responsibilities for the functions of families" (Vanier Institute for the Family)

Family form—the number of adults, the nature of the relationship between the adults, and the number of generations

Fashion—the prevailing style; current custom in dress

Feedback—part of the communication process; actions and verbal responses that indicate one is listening, understanding, and encouraging the communication to continue

Financial institutions—banks, trust companies, credit unions and other institutions that offer a variety of financial services

Fixed expenses—items on which one regularly spends money

Flexible expenses—items on which one occasionally spends money

Food group—one category of foods in Canada's Food Guide in which all the items have similar nutritional value; for example, milk products

Food-borne illnesses—sicknesses contracted from contaminated or diseased foods

Foster family—a family that takes temporary care of children until they can return to their biological family

Functional relationships—relationships that meet one's academic, financial, health, or other needs, rather than emotional needs

Functions of the family—physical maintenance and care of family members; addition of new family members through procreation or adoption; socialization of children; production, consumption, and distribution of goods and services; nurturance and love (Shirley Zimmerman)

Goal—something a person wants to achieve at a certain time in his or her life

Goods and services—things for sale, wares and arrangements for finding something useful or necessary

Governments—systems of ruling; rules or authorities over a country, province, district, etc.

Growth—physical changes that occur in the body

Guarantee—promise or pledge to replace goods if they are not as represented

Harassment—unwelcome and uninvited attention that causes the victim distress

Heredity—traits inherited from the unique gene set of one's parents

Home—the place where a person or family lives; one's own house

Homeless—having no permanent form of housing and living either on the street or in temporary or emergency shelters

Household—a home and its affairs

Human resources—the skills, talents, and qualities of people that can help an individual or an organization to achieve a goal

I-messages—non-threatening, non-blaming messages of clear and assertive communication

Impulse buying—buying a product, etc. with little or no planning

Income—any money that one takes in that would be available for spending

Infertile—not fertile, sterile

Infomercial—a television program that consists entirely of long advertisements

Insurance—a form of protection that can be purchased from insurance companies; by paying a set amount on a regular basis, one can be assured of receiving money in the event of a loss

Interest—a fee paid for the use of one's money

Interruptions—occurrences that cause a break in concentration; hinder; obstruct

Irreconcilable differences—grounds for divorce for couples who can no longer agree

Laundry—clothes, etc., washed or to be washed; the washing and ironing of clothes, etc.

Life expectancy—the number of years one can reasonably expect to live

Life span—the length of time that a person lives

Lifestyles—the typical habits, pastimes, attitudes, standard of living, etc. of a person or group

Limits—boundaries

Lone-parent or single-parent family—a family that consists of one parent and one or more biological or adopted children

Management process—taking control of the factors that affect your life, enabling you to get what you want by setting goals and using the resources available to you

Manners—ways of behaving towards others; polite behaviour

Mediation—the process of settling conflict with the help of an unbiased third party

Monogamous—having only one spouse or sexual partner at a time

Multitasking—doing more than one thing at a time

Natural environment—the earth as natural surroundings; the air, water, land around us

Needs—those things that are necessary for a person's growth and development

Negotiate—talk over and arrange terms; parley; confer; consult

Negotiation—the process of settling conflict through problem solving and cooperation

Non-human resources—tangible items such as money, tools, sources of information, or other materials, such as personal possessions that can help an individual or an organization achieve a goal

Non-renewable resources—natural resources that cannot be replaced, such as coal or oil

Non-verbal communication—conveying a message using elements other than words; elements include gestures, eye contact, body movements, facial expressions

Nuclear family—a family that consists of two parents and one or more children living together

Nurture—the act or process of raising or rearing; training, education, nourishment; the sum of

environmental factors acting on a person, as opposed to genetic makeup

Nutrients—chemical substances that make up food and provide the essential nourishment for the body to function, grow, repair itself, and produce energy

Nutrition—food; nourishment

Opinions—ideas based on individual observations or beliefs

Parenting style—parents' approach to raising and caring for their children

Passive listening—hearing a message without really taking in the meaning

Patriarchal—ruled by the father in a family or tribe

Peer pressure—influence on one's behaviour by others in a similar age or interest group

Peers—people of the same interest or age group with whom one shares similarities

Personal responsibility—a person's reliability and accountability for the decisions he or she makes

Polygamous—having more than one spouse or sexual partner at a time

Power—strength; might; force; the ability to control the behaviour of another

Prejudice—a type of bias against an individual or group that has certain characteristics

Principles of design—balance, proportion, and harmony

Priorities—the most important things that need to be done

Problem solving—the process of settling conflict or overcoming a difficulty

Procrastination—putting things off

Protection—keeping from harm; defence

Psychology—the study of how people think, feel, and are motivated

Public services—services funded by tax dollars and available to the public (e.g., Health care, education, fire and police protection)

Rapport—a harmonious or agreeable relationship or connection

Recipes—sets of directions for preparing something to eat or drink

Reciprocity—a mutual exchange; giving and receiving

Recycling—the reprocessing of waste material so that it can be used again or any other reuse of something

Relationships—the bonds formed between people based on common interests, and often, on affection

Renewable resources—natural resources that can be replaced, such as solar power

Research question—the basis of an investigation; states what the researcher wants to learn and suggests how the research will be conducted

Resources—any means of reaching a goal

Respect—honour; esteem

Responsibility—an action or task for which one is held accountable

Rights—just claims, titles, or privileges; usually defined by laws and charters

Role model—someone regarded by others as an inspiration for younger or less experienced people to imitate

Roles—the parts one plays in real life

Safety—freedom from harm or danger

Schedule—a way of planning time on a daily basis

Self-awareness—being attuned to what one believes and feels

Self-concept—a person's perception of himself or herself based on attitudes and feelings; also called self-image or identity

Self-discipline—controlling one's own behaviour

Self-esteem—a person's judgement of his or her self-concept or self-image

Shelter—protection, refuge

Sibling rivalry—the competition arising from jealousy between brothers and sisters

Siblings—brothers and sisters

Skill—an ability that one has learned, such as riding a bicycle or using a computer

Smoke detectors—devices designed to detect the presence of smoke

Social policy—a course or method of social action that has been deliberately chosen and that guides or influences future decisions

Social sciences—the study of human behaviour; includes political science, psychology, sociology, anthropology, and economics

Socialization—teaching the skills, values, and attitudes needed to be prepared for life; make fit for living with others

Sociology—the study of how people behave with others within families, groups, and society

Spouse—husband or wife; partner

Standard of living—the degree of material comfort available to a person, class, or community

Standards—anything taken as a basis of comparison; model

Statisticians—workers who develop and apply mathematical or statistical techniques to solve problems in fields such as physical and biological science, engineering, social science, business, and economics

Stepparents—the parents of children in blended families; a mother's or father's later spouse

Talent—an ability that comes naturally, such as the ability to sing in perfect pitch

Taxation—money paid by the people for the support of the government, for public works, etc.

Technology—a human-made resource that combines scientific knowledge with natural resources to create tools that extend one's physical or mental abilities

Temperament—how a person responds to experiences within his or her environment, both positive and negative

Time management—the process of managing one's time to accomplish goals in a timely manner and to ensure that all the things one wants to do get done

Tolerance—respect for the right of others to have different beliefs and interests than one's own

Transfer payments—payments made by a government to an individual, in the form of benefits, or to a lower level of government

Values—the ideas and beliefs that guide a person's life

Verbal communication—conveying a message using words

Volunteering—working without being paid for your time

Wants—those things that will make a person's life more pleasant

Wardrobe—a stock of clothes

Warranty—a manufacturer's written guarantee regarding the extent to which defective goods will be repaired or replaced

Index

Credits

Every effort has been made to find and to acknowledge correctly the sources of the material reproduced in this book. The publisher welcomes any information that will enable him to rectify, in subsequent editions, any errors or omissions.

Chapter openers
left to right: PhotoDisc; PhotoDisc; PhotoDisc; PhotoDisc; PhotoDisc; Eyewire; PhotoDisc; PhotoDisc; Eyewire; PhotoDisc; Eyewire; Eyewire; SuperStock; PhotoDisc

Chapter 1
Page 6, Irwin Publishing; **page 7**, Reprinted with special permission of King Features Syndicate; **page 9**, John Henley/Firstlight.ca

Chapter 2
Page 14 Alan McArthur TPL/Firslight.ca; **page 15** Zigy Kaluzny/Stone; **page 17** PhotoDisc; **page 18** Tom Stewart/Firstlight.ca; **page 19** PhotoDisc; **page 21** (top) David DeLossy/Image Bank; **page 21** (bottom) Eyewire; **page 22** Amwell/Stone; **page 24** (top) PhotoDisc; **page 24** (bottom) PhotoDisc; **page 26** (top) Ian O'Leary/Stone; **page 26** (bottom) Miro Vintoniv/Image Network Inc.; **page 27** Mug Shots/Firstlight.ca

Chapter 3
Page 33; John Lehman/CP Picture Archive; **page 33** PEANUTS reprinted by permission of United Feature Syndicate, Inc.; **page 34** photo(t-l) PhotoDisc, page clockwise, DiMaggio/Kalish/Firstlight.ca; Jason Homa/Image Bank; Lambert/Image Bank; **page 36** Reprinted by permission of Universal Press Syndicate; **page 39** (l-r) Image Network Inc.; Michael Melford/Image Bank; PhotoDisc; Joe Luis Pelaez Inc./Firstlight.ca; PhotoDisc; **page 40** Eyewire; **page 43** Miao China Tourism Press, Wang/Image Bank; David Young-Wolff/Stone; **page 44** PhotoDisc

Chapter 4
Page 49 David Stoecklein/Firstlight.ca; **page 51** Irwin Publishing; **page 55** Eyewire; Justin Pumfrey/Firstlight.ca; **page 56** Lou Jones/Image Bank; **page 57** Juan Alvarez/Image Bank; **page 57** PEANUTS reprinted by permission of United Feature Syndicate, Inc.

Chapter 5
Page 65 Bard Martin/Image Bank; **page 66** PhotoDisc; Copyright Tribune Media Services, Inc. All Rights Reserved. Reprinted with permission; **page 67** Irwin Publishing; Ron Sherman/Stone; **page 68** Reprinted with special permission of King Features Syndicate;

page 69 Chips TPL/Firstlight.ca; **page 70** Arthur Tilley/Stone; Eyewire; **page 71** ©Lynn Johnston Productions, Inc./Distributed by United Feature Syndicate, Inc.; **page 72** Paul Chiasson/CP Picture Archive; **page 72** Copyright Tribune Media Services, Inc. All Rights Reserved. Reprinted with permission; **page 73** David Young-Wolff/Stone; **page 74** Zigy Kalusky/Stone; Ian Shaw/Stone; **page 75** Carol Kohen/Image Bank

Chapter 6
Page 80 PhotoDisc; **page 81** Reprinted with special permission of King Features Syndicate; Ronnie Kaufman/Firstlight.ca; **page 82** Eric Larrayadieu/Stone; ©Lynn Johnston Productions, Inc./Distributed by United Feature Syndicate, Inc.; **page 84** PhotoDisc; **page 87** PhotoDisc; **page 89** ©Lynn Johnston Productions, Inc./Distributed by United Feature Syndicate, Inc.; **page 91** Juan Silva Productions/Image Bank; **page 91** Rob Lewine/Firstlight.ca; **page 92** PhotoDisc; Eyewire; Eyewire; **page 93** Reprinted with special permission of King Features Syndicate; Eyewire

Chapter 7
Page 98 Reprinted with special permission of King Features Syndicate; **page 99** PhotoDisc; **page 100** PhotoDisc; **page 102** PhotoDisc; **page 104** Reprinted with special permission of King Features Syndicate; **page 107** PhotoDisc; **page 109** Ben Walsh/Firstlight.ca; **page 110** SuperStock; **page 111** Kids Help Phone

Chapter 8
Page 116 David deLossy/Image Bank; **page 119** Serge Attal/Image Bank; Charles Gupton/Firstlight.ca; **page 120** Romily Lockyer/Image Bank; **page 121** Paul Chesley/Stone; **page 123** Carol Kohen/Image Bank; **page 125** PhotoDisc; Eyewire; Jurgen Magg/Firstlight.ca; Mark Richards/PhotoEdit

Chapter 9
Page 131 Thomas Mower Martin, Encampment of Woodland Indians, 1880, Glenbow Collection, Calgary, Canada; Art Wolfe/Stone; **page 136** CP Picture Archive, **page 133** Courtesy of CIDA; **page 134** Glenbow Archives, Calgary, Canada NA-1982-2; **page 135** McCord Museum, MP.1976.24.; Lambert03RLAM/Image Bank; **page 137** PhotoDisc; Glenbow Archives, Calgary, Canada NA-5501-3;

page 138 Glenbow Archives, Calgary, Canada
NA-1752-29; **page 140** Claire Matches/Firstlight.ca;
page 141 ©Lynn Johnston Productions, Inc./
Distributed by United Feature Syndicate, Inc.

Chapter 10
Page 147 PhotoDisc; PhotoDisc; **page 149**
SuperStock; PhotoDisc; **page 150** Eyewire; **page 151**
Jose Luis Pelaez Inc.; **page 155** Eyewire; **page 157**
Elizabeth Hathon/Firstlight.ca; **page 159** PhotoDisc;
page 160 PhotoDisc; Eyewire; **page 162** Eyewire

Chapter 11
Page 166 Lambert 03RLAM/Image Bank; **page 167** Zigy
Kaluzny/Stone; Eyewire; Reprinted with special permission of King Features Syndicate; **page 171** Reprinted
with special permission of King Features Syndicate;
page 174 PhotoDisc, **page 175** Eyewire; **page 176**
PhotoDisc; **page 178** ©Lynn Johnston Productions,
Inc./Distributed by United Feature Syndicate, Inc.;
page 180 Courtesy of WSIB; Michael Rosenfeld/Stone;
page 182 Vic Bider/Image Network Inc.

Chapter 12
Page 187 Mad Cow Studio/Firstlight.ca; **page 188**,
clockwise, David Young-Wolff/PhotoEdit; Michael
Newman/PhotoEdit; Images/Firstlight.ca; **page 189**
PhotoEdit; G & M David de Lossy/Image Bank;
page 191 David Young-Wolff/PhotoEdit; Dana White/
PhotoEdit; **page 193** Jonathan Hayward/CP Picture
Archive; David Kelly Crow/PhotoEdit ; **page 196**
Phill Snel/CP Picture Archive

Chapter 13
Page 203 Andreas Pollock/Stone; Eyewire; Walter
Biskow/Image Bank; David DeLossy/Image Bank;
page 208 ©Lynn Johnston Productions, Inc./
Distributed by United Feature Syndicate, Inc.;
page 211 Eyewire; **page 213** Eyewire; **page 215**
Image Network Inc.; Yellow Dog Productions/Image
Bank; Paul Chiasson/CP Picture Archive

Chapter 14
Page 220 PhotoDisc; Juan Silva Productions/Image
Bank; Eyewire; **page 223** David Young-Wolff/
PhotoEdit; **page 224** PhotoDisc; **page 226** PhotoDisc;
page 228 Eyewire; **page 229** Jon Riley/Stone; Juan
Silva Productions/Image Bank;

Chapter 15
Page 234 CP Picture Archive (Ryan Remiorz);
page 235 David Young-Wolff/PhotoEdit; **page 237**
Dan Callis/CP Picture Archive; Eyewire; Eyewire;
CP Picture Archive (Fred Lum); PhotoDisc; **page 238**
Don Smetzer/Stone; **page 239** Images/Firstlight.ca;
David Rossiter/CP Picture Archive; FirstLight.ca;
Dana White/PhotoEdit; **page 240** Andrew
Vaughan/CP Picture Archive

Chapter 16
Page 246 Reprinted with special permission of King
Features Syndicate; **page 247** Image Bank; **page 248**
Bob Thomas/Stone; Sean Justice/Image Bank; **page 249**
Eyewire; **page 250** O Claire Photex/FirstLight; **page 251**
©Lynn Johnston Productions, Inc./ Distributed by
United Feature Syndicate, Inc.; **page 252** PhotoEdit;
page 253 Britt Erlanson/Image Bank; **page 254**
PhotoDisc; **page 256** ©Lynn Johnston Productions,
Inc./Distributed by United Feature Syndicate, Inc.;
page 259 Lindy Powers/Image Network Inc.; **page 260**
Mary Kate Denny/PhotoEdit; **page 262** Inc. Janeart/
Image Bank; **page 263** Britt Erlanson/Image Bank

Chapter 17
Page 269 David Young-Wolff/PhotoEdit; **page 273**
Michael Newman/PhotoEdit; **page 274** Reprinted
with special permission of King Features Syndicate;
page 275 Firstlight.ca; **page 276** Michael Newman/
PhotoEdit; Spencer Grant/PhotoEdit; **page 278** David
Young-Wolff/PhotoEdit; **page 279** Eric Larrayadieu/
Stone; Paddy S'broker/Firstlight.ca; **page 279** Rita
Maas/Image Bank; Rita Maas/Image Bank; **page 282**
Michael Rosenfeld/Stone; Reprinted with special
permission of King Features Syndicate; **page 283**
Randy M. Vry/Firstlight.ca

Chapter 18
Page 289 Tracy Frankel/Image Bank; **page 290** Kevin
Frayer/CP Picture Archive; Elise Amendola/CP
Picture Archive; **page 291** Kevin Frayer/CP Picture
Archive; **page 294** PhotoDisc; Alan Becker/Image
Bank; **page 295** Ian Shaw/Stone; **page 296** Eyewire;
page 298 B.Harris/Firstlight.ca; **page 300** Image
Network Inc.; **page 302** ©Lynn Johnston Productions,
Inc./Distributed by United Feature Syndicate, Inc.;
page 304 Image Network Inc.; **page 305** Leighton
Image; **page 306** Irwin Publishing; **page 307** Sean
Connor/Edmonton Sun; Sean Connor/Edmonton Sun

Chapter 19
Page 312 Laurence Dutton/Stone; **page 314**
PhotoDisc; **page 315** Richard Berenholz/Firstlight.ca;
PhotoDisc; **page 316** PhotoDisc; **page 317** PhotoDisc;
page David Young Wolff/Stone; Premium Stock/
Firstlight.ca; Premium Stock/Firstlight.ca; **page 320**
Tony Freeman/PhotoEdit; **page 321** Tony Freeman/
PhotoEdit; Superstock; **page 322** PhotoDisc; **page 326**
PhotoDisc; **page 328** PhotoDisc; T. Stewart/Firstlight.ca

Chapter 20
Page 350 ©Lynn Johnston Productions,
Inc./Distributed by United Feature Syndicate, Inc.;
page 351 Stone; **page 353** Firstlight.ca; **page 355**
SW Production/Image Network Inc.; **page 357** Jon
Riley/Stone; **page 358** David Epperson/Stone